ENVIRONMENTAL REMEDIATION TECHNOLOGIES, REGULATIONS AND SAFETY

HUMIDITY SENSORS: TYPES, NANOMATERIALS AND ENVIRONMENTAL MONITORING

Environmental Remediation Technologies, Regulations and Safety

Additional books in this series can be found on Nova's website under the Series tab.

Additional E-books in this series can be found on Nova's website under the E-books tab.

ENVIRONMENTAL REMEDIATION TECHNOLOGIES, REGULATIONS AND SAFETY

HUMIDITY SENSORS: TYPES, NANOMATERIALS AND ENVIRONMENTAL MONITORING

CHRISTOPHER T. OKADA
EDITOR

Nova Science Publishers, Inc.
New York

For permission to use material from this book please contact us:
Telephone 631-231-7269; Fax 631-231-8175
Web Site: http://www.novapublishers.com

LIBRARY OF CONGRESS CATALOGING-IN-PUBLICATION DATA

Humidity sensors : types, nanomaterials, and environmental monitoring /
edited by Christopher T. Okada.
p. cm.
Includes bibliographical references and index.
ISBN 978-1-61209-246-1 (hardcover)
1. Hygrometers. I. Okada, Christopher T.
QC916.H86 2011
551.57'10287--dc22

2010047633

Published by Nova Science Publishers, Inc. † New York

CONTENTS

PREFACE

There is a substantial interest in the development of relative humidity sensors for applications in monitoring moisture levels at home, in clean rooms, cryogenic processes, medical and food science. This new book presents current research in the study of humidity sensors, including measuring methods and standards of water vapor sorption and humidity; environmental and bio-medical applications using quartz crystal microbalance humidity sensors and surface modified electrospun nanofibrous membranes for humidity detection.

Chapter 1 - Under environmental conditions water exists in all three classical states of matter: solid, liquid and gas. The water molecule is non-linear and therefore polar. In comparison with other liquids water has anomalous features; about 63 exceptional properties are recorded. This article starts with reviewing properties of water, typical occurrences and definitions such as relative and absolute humidity and moisture content.

Water is present everywhere in nature and engineering; it may be helpful or harmful. The survey concerns both: atmospheric hygrometry and usual measuring methods of the moisture content of solids and liquids as well as water sorption. The determination of the atmospheric humidity is among the more difficult problems in metrology. In contrast, humidity determination of materials is simple; however the definition of the dry state is difficult. Because water is bound at and in solids and liquids in many different ways it turns out that the humidity content of materials is difficult to define and to measure accurately. We provide a survey on the measuring methods, describe the most important ones and discuss advantages and accuracy.

In the search for extraterrestrial water special remote measuring methods have been developed analysing the spectrum of electromagnetic radiation either of natural sources or produced by a probe and reflected. Spacious deposits have been detected photographically. In situ investigations are made using conventional methods.

With regard to the problems of measurements, standardisation of measuring methods and procedures is required. There exist many institutions, which are engaged in investigating the use of water and standardising measuring methods. In tabular form, we give a survey on existing standards.

Chapter 2 - Nano-assembled thin films prepared by a layer-by-layer approach on QCM resonators were used as sensitive elements for monitoring relative humidity. Different types of films, polymer-based $(PDDA/PSS)_n$ film and porphyrin-based $(PDDA/TSPP)_n$ and $(PDDA/MnTSPP)_n$ films (where n=5, 10, and 15), showed linear responses to relative humidity changes in the range of 4%–94%. An increase of the number of bilayers deposited on the QCM electrode enhanced the sensitivity to relative humidity. The porphyrin-based films showed ca. 4.5 times higher sensitivity towards relative humidity than the polymer-based films, owing to the presence of the higher number of free sulfonic acid groups. Incorporation of the metal ion into the porphyrin pyrrole ring was not significant in sensitivity. However, all films showed a good reversibility to the stepwise changes of relative humidity and the response and recovery times (t_{90}) were very fast, within 15 sec.

Additional examples of the QCM humidity sensors for environmental and medical sensing are presented. In particular, a respiratory monitoring system based on QCM frequency oscillations was designed and investigated. The developed system, in which the sensor response reflects lung movements, was able to track human breathing parameters such as respiratory rate, tidal volume, and inspiratory/expiratory flow, which were tested on healthy volunteers.

Chapter 3 - This chapter presents holographic devices for sensing the relative humidity in the environment. The principle of operation of two main types of holographic sensors based on transmission and reflection holograms is described and their properties are theoretically modelled. The effect of different factors such as change of the overall refractive index, change of the hologram's thickness due to swelling/shrinkage of the layer and change in the refractive index modulation due to absorption/desorption of moisture are considered. In the experimental studies emphasis is given to reflection holographic humidity sensors which can provide a visual indication of the relative humidity observed as a change of the colour of the reflected light. The work presented focuses on holograms recorded in acrylamide based photopolymers. The results from studies of their sensitivity, reversibility, relative humidity range of operation and response time are summarised. The operational temperature range is also studied.

The potential use of the humidity sensors as user-verifiable security holograms is described as well as applications in packaging and environmental sensing.

Chapter 4 - Electrospinning exhibits the unique ability to produce diverse forms of fibrous assemblies with remarkable specific surface area and high porosity, making electrospun nanomaterials highly attractive to ultra-sensitive sensors and other nanotechnological applications. In this chapter, electrospun fibrous chitosan (CS) membranes modified with polyelectrolytes (polyacrylic acid (PAA) and polyethyleneimine (PEI)) were used as sensing materials coated on quartz crystal microbalance (QCM) for humidity detection. A series of nanofibrous membranes containing polyelectrolytes are fabricated on QCM and characterized regarding their morphology, sensitivity and hysteresis. Sensing experiments are carried out by measuring the resonance frequency shifts of QCM due to the additional mass loading. The results indicated that the response of the sensors increased by three to four orders of magnitude with increasing the relative humidity (RH) from 8 to 95% at room temperature, exhibiting high sensitivity and that, in the range of 20-95% RH, the $Log(\Delta f)$ showed a good linearity. Additionally, the prepared sensors exhibited remarkably enhanced sensitivity with increasing the coating loads of polyelectrolytes on QCM. Furthermore, the dynamic Langmuir adsorption model was used to determine the kinetic parameters such as adsorption rates and Gibbs free energy for various RH between 70 and 95%.

Chapter 5 - Humidity is a dynamic parameter that is essential for various fields of industry as well as human lives. There is a substantial interest in the development of relative humidity (RH) sensors for applications in monitoring moisture level at home, in clean rooms, cryogenic processes, medical and food science, and so on. Humidity sensors based on changes in the capacitance or resistance of the sensing element from absorption of water vapor have been investigated extensively. Alternatively, surface acoustic wave (SAW) resonant sensors have also been developed with polymer films deposited on top of a SAW resonator as the sensing layer. In this chapter, a novel RH sensing device using ZnO based film bulk acoustic-wave resonator (FBAR) was described. The resonant frequency of the FBAR decreased in a two-stage manner as RH increased in the environment. For low RH (RH < 50%), a frequency downshift of 2.2 kHz per 1% RH change was observed. This effect was attributed to water molecules replacing the adsorbed oxygen on the ZnO surface, thus increasing the density of the film, resulting in a frequency drop. For high RH

(RH > 50%), a frequency downshift of 8.5 kHz per 1% RH change was obtained, which was due to the mass loading effect of the water layers formed on the ZnO surface. It was demonstrated that the two-stage response of the FBAR can be interpreted using the power law theory for semiconductor gas sensors and the mass loading effect of the resonator. UV light was applied to monitor its effects on the humidity sensing performance of the FBAR. UV can enhance the sensitivity at low RH (frequency downshift increased to 3.4 kHz per 1% RH change), while degrading the sensitivity at high RH (frequency downshift decreased to 5.7 kHz per 1% RH change). The mechanism of the influence of the UV illumination was interpreted.

In: Humidity Sensors
Editor: Christopher T. Okada

ISBN 978-1-61209-246-1

Chapter 1

WATER VAPOUR SORPTION AND HUMIDITY – A SURVEY ON MEASURING METHODS AND STANDARDS

Erich Robens[1], Katrin Rübner[2], Peter Klobes[3] and Devrim Balköse[4]

[1]Institut für Anorganische Chemie und Analytische Chemie der Johannes Gutenberg-Universität, Duesbergweg 10-14, D-55099 Mainz, Germany,
[2]BAM - Federal Institute for Materials Research and Testing, Unter den Eichen 87, D-12205 Berlin, Germany,
[3]BAM - Federal Institute for Materials Research and Testing , Richard-Willstätter-Str. 11, D-12489 Berlin, Germany
[4]Izmir Institute of Technology Faculty of Engineering Department of Chemical Engineering, Gulbahçeköyü Urla İzmir Turkey

ABSTRACT

Under environmental conditions water exists in all three classical states of matter: solid, liquid and gas. The water molecule is non-linear

1 erich.robens@t-online.de
2 katrin.ruebner@bam.de.
3 peter.klobes@bam.de
4 devrimbalkose@iyte.edu.tr

and therefore polar. In comparison with other liquids water has anomalous features; about 63 exceptional properties are recorded. This article starts with reviewing properties of water, typical occurrences and definitions such as relative and absolute humidity and moisture content.

Water is present everywhere in nature and engineering; it may be helpful or harmful. The survey concerns both: atmospheric hygrometry and usual measuring methods of the moisture content of solids and liquids as well as water sorption. The determination of the atmospheric humidity is among the more difficult problems in metrology. In contrast, humidity determination of materials is simple; however the definition of the dry state is difficult. Because water is bound at and in solids and liquids in many different ways it turns out that the humidity content of materials is difficult to define and to measure accurately. We provide a survey on the measuring methods, describe the most important ones and discuss advantages and accuracy.

In the search for extraterrestrial water special remote measuring methods have been developed analysing the spectrum of electromagnetic radiation either of natural sources or produced by a probe and reflected. Spacious deposits have been detected photographically. In situ investigations are made using conventional methods.

With regard to the problems of measurements, standardisation of measuring methods and procedures is required. There exist many institutions, which are engaged in investigating the use of water and standardising measuring methods. In tabular form, we give a survey on existing standards.

1. Introduction

Water is present everywhere in nature; this could be advantageous or disadvantageous. Handling of a variety of materials includes often the use of water, as solvent or dispersant medium, adsorptive, reacting agent, cleaning agent and lubricant. For handling, storage and quality assurance of foods, pharmaceuticals, textiles, plastics, paper, wood and building materials the water content must be known. Many processes, such as corrosion protection, refrigeration, air conditioning in museums, greenhouses, saunas, laboratories and climate chambers as well as humidors for manufacturing of pure materials, require the control of the air humidity. An extreme control of humidity is required for insulating gases, such as those in transformers and power plants, as well as protecting and other pure bulk gases. Natural gas transport requires control of humidity down to a measuring range of ppm in presence of high contents of solid and liquid contaminants as well as corrosives in varying concentrations. In general, insulating liquids must be free from traces of water.

Very different applications of humidity measurements are made under very different circumstances within a wide measuring range from nearly zero up to saturation and sometimes with high sensitivity. Thus, a great variety of measuring methods is applied [1-4]. Because the term "humidity" is not well defined and because the different measuring methods deliver not always comparable results standardisation of measuring methods and procedures is required.

A humidity measurement covers two groups of measuring problems: the atmospheric humidity and the material moisture. In both groups, the phenomena and its mathematical description are very different and so are the methods and instrumentation for measurement and control. The two groups are of high importance for life and techniques and thus, a large variety of measuring instruments is available on the market.

2. Water Structures

Under global environmental conditions water exists in all three classical states of matter: solid, liquid and gas. The water molecule is non-linear (Figure 1) and therefore polar. In comparison with other liquids it has anomalous features; about 63 exceptional properties are recorded [5].

Water in the Gaseous Phase

In gaseous phase water is present as a dimeric complex along with single molecules [6] (Figure 2). Air contains always water vapour in a concentration from near zero up to the saturation value determined by temperature. In addition, liquid or solid agglomerates may be present and visible as clouds, droplets, snow or hailstones. Contaminations should be taken into account, thus, little droplets of ionic solutions are formed with salt from sea water, and dust provides nuclei for condensation.

Water in Liquids

Liquid water consists mostly of aggregates of several molecules (Figure

3). In addition, water can form mixtures and solutions in liquids. Due to its high polarity, dissociation and ionization of substances can occur.

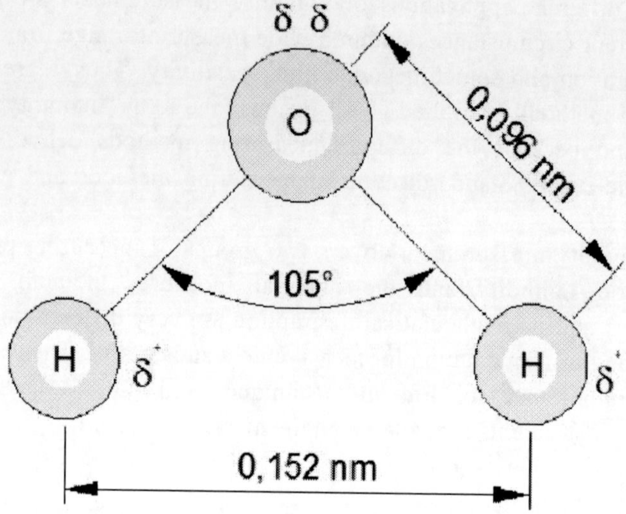

Figure 1. Schematic structure of a water molecule.

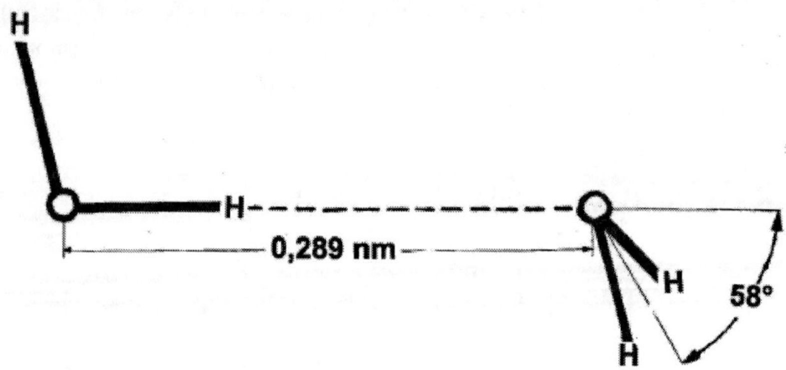

Figure 2. Structure of dimeric water molecule in vapour.

Water Near Solid Surfaces

At ambient temperatures up to about 35 °C, free bulk water consists of unstable and quickly varying aggregates of several molecules (Figure 3);

single molecules are seldom. As shown in Table 1, water is bound at a solid differently.

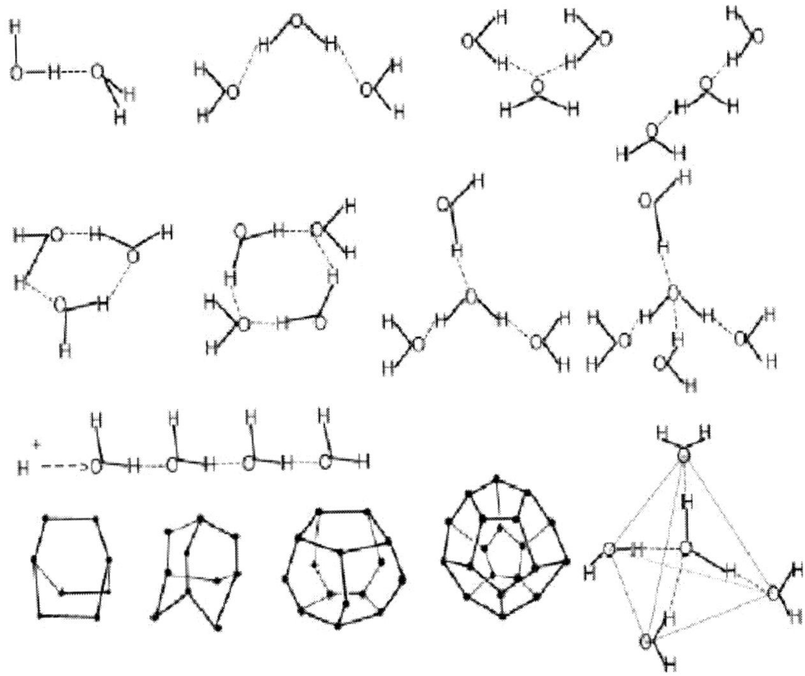

Figure 3. Possible structures of water molecules.

Interpacked water filling larger pores and spaces between grains has the same structure as free bulk water. Likewise, capillary water fills capillaries (macropores) and nanopores (mesopores). Due to water uptake, the structure of some materials can store large amounts of water and/or can undergo swelling. A solid surface exposed to the atmosphere is more or less covered with physisorbed water molecules attracted by van der Waals forces (London dispersion forces). These are induced, fluctuating dipoles. Whilst their temporary mean value is zero, the binding energy per atom is of the order of 0.5 eV. Down to temperatures of -50 °C, liquid water layers have been observed in fissures of permafrost structures. About two layers of physisorbed water molecules at a solid surface or at the surface of ice behave in a quasi-liquid state down to a temperature of 33 K. Such quasi-liquid layers behave like free water; however, the surface influences the structures within about eight layers. Thus, bound water has properties different from those of free water [6]. Physisorption is reversible because the electron orbital structure is not influenced and only weak deformation of the molecular grid occurs. Parts of the adsorbate vaporise with decreasing vapour pressure or increasing temperature and a new equilibrium responds to the new ambient state. Nevertheless, to remove physisorbed water completely in a reasonable time, warming up is often required in addition to vacuum.

Binding forces at polar surfaces and induced electrostatic dipoles at metallic surfaces are much stronger; they are of the order of 0.1 kJ mol^{-1}. Though there is still no chemical bond, and thus, such bindings should be classified as physisorption.

Chemisorption is a dissociative stoichiometric bonding at surface molecules of the solid. Binding enthalpies range between 100 and 500 kJ mol^{-1} Chemisorption is irreversible. Energy is required to release molecules, whereby hydroxyl groups may recombine to form water molecules.

Absorptive binding of water is associated with diffusion of water molecules, which are first physisorbed at the surface, into free molecular sites in the interior of a material.

The formation of a hydrate and binding as water of crystallisation are chemical reactions.

Table 1. Water near solid surfaces

Type of water	Formation process
Free bulk water - volume water - interpacked water	Filling of volumes Filling of macropores and interspaces between particles
Water bound physically at the solid surface - physisorbed water layers -capillary water in mesopores	Adsorption (physisorption) at the outer surface and at pore walls and micropore filling Adsorption (physisorption) and condensation in mesopores (capillary condensation)
Water bound chemically at the surface - chemisorbed water	Adsorption and chemisorption at the outer surface and at pore walls
Water bound in the solid bulk material - absorbed water	Adsorption and diffusion → absorption

Binding enthalpies are stronger than those of physisorption. Nevertheless such reactions may be completely reversible and may take place even at low temperatures.

Natural and technical surfaces are always contaminated. Impurities at the surface are hardly avoidable. Water may emulsify or dissolve such species by forming an ionic salt solution e.g.

A typical effect accompanying water sorption is swelling. By expanding the matrix of dry material, additional volume is available for water molecules. For example, the small specific surface area of dry clay minerals increases by water uptake to values between 10 and 2200 $m^2 g^{-1}$ as determined by the BET method.

3. Air and Gas Humidity

The following chapter is concerned with the water content of air and process gases [7]. The earliest and still widely employed application of detection of air humidity is in weather forecast because humidity is one important parameter. Weather forecast has been made most probably since the beginning of mankind. Animals and plants are equipped with humidity sensors as well, and it is believed that some animals can forecast weather.

Although living in highly developed countries and mostly in protected environments, weather is still governing our life and it influences also industrial processes. Water vapour in air causes atmospheric phenomena such as rain, snow or frost and affects solids by corrosion of metals, caking of powders and deterioration of delicate technical devices. Humidity sensors with linear, quick and reversible responses have been developed for measuring water vapour concentration in air [8].

Historical

The first record of a water vapour adsorption experiment is found in the Bible. In the Book of Judges [9] we read,

> "And Gideon said to God: If you want me to save Israel by my hand, as you said, behold, I put fresh sheared wool on the floor and if the dew falls only on the wool, and it will be dry at all the earth beside, than shall I know that you want to save Israel by my hand, as you have said. And it was so:

When he got there in the early morning and wrung out the wool, he could press out the dew from the wool, a bowl full of water. After that Gideon said to God: Your anger should not flare out against me, if I speak again: Only once more I will try it with the wool: Only the wool shall be kept dry, and dew shall be upon all the ground. And God did so in the following night: The wool remained dry and the dew was on all the ground."

Already Luther remarked that the description of the experiment is unclear. Nevertheless, there are plausible explanations for the inconsistent results [10-11].

A strange volumetric device, which was in use until recently, should be mentioned: "The doctor of Cairo" (Figure 4). These artistic bowls served for dew sampling during cool nights. The condensed water was used for medical purposes.

Figure 4. “The Doctor of Cairo” (dew sampler).

The aim of the first gravimetric adsorption measurements was to forecast weather by determination of atmospheric humidity. The very first to describe such an instrument [12], was the German cardinal Nicolaus Cusanus (1401-1463). In his book "Idiota de Staticis Experimentis", which was published in 1450, that means "The layman about experiments with a balance" (Figure 5) [13], he let this ignorant person, probably his mechanic, suppose,

"If anyone hangs on one side of a big balance with dry wool and loads of stone on the other side until equilibrium is established, at a place and in air of moderate temperature he could observe that with increasing humidity the weight of the wool increases and with increasing dryness of the air it decreases. By these differences it is possible to weigh the air and it is likely that one might perform weather forecasting."

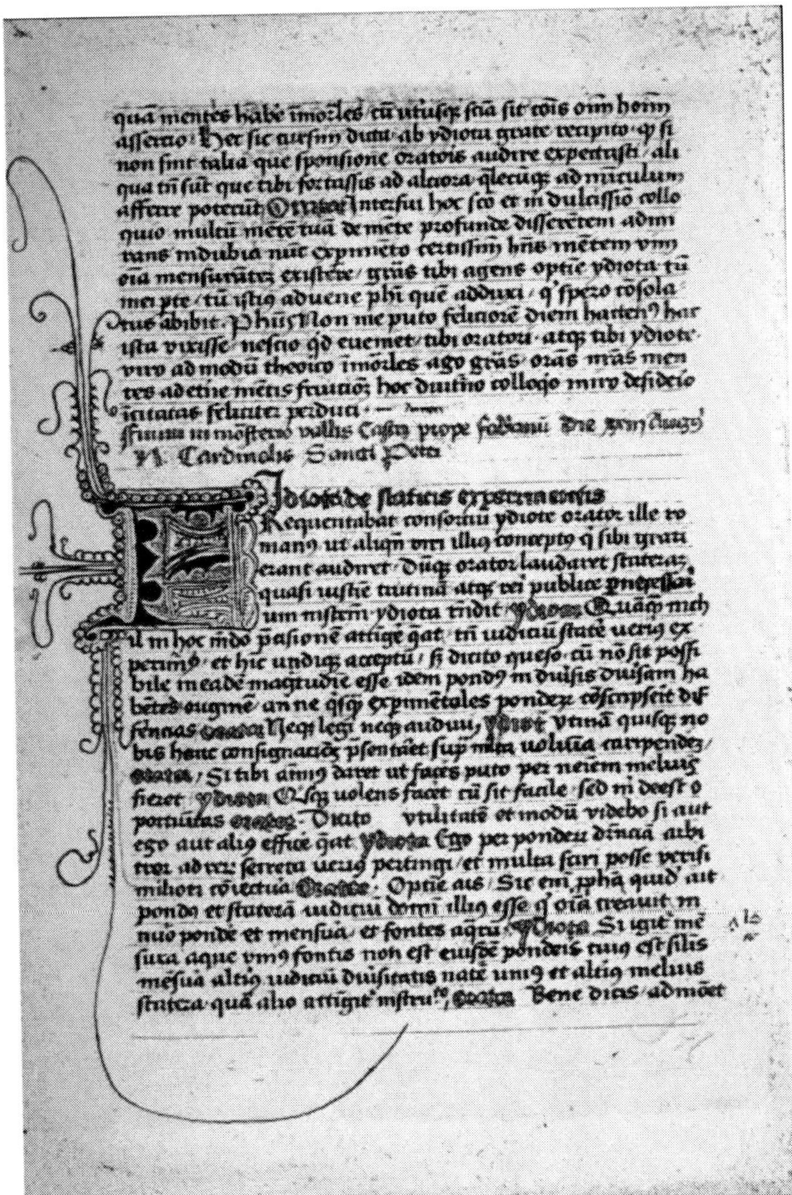

Figure 5. Nicolaus Cusanus (1401-1463). "Idiota de Staticis Experimentis" 1450.

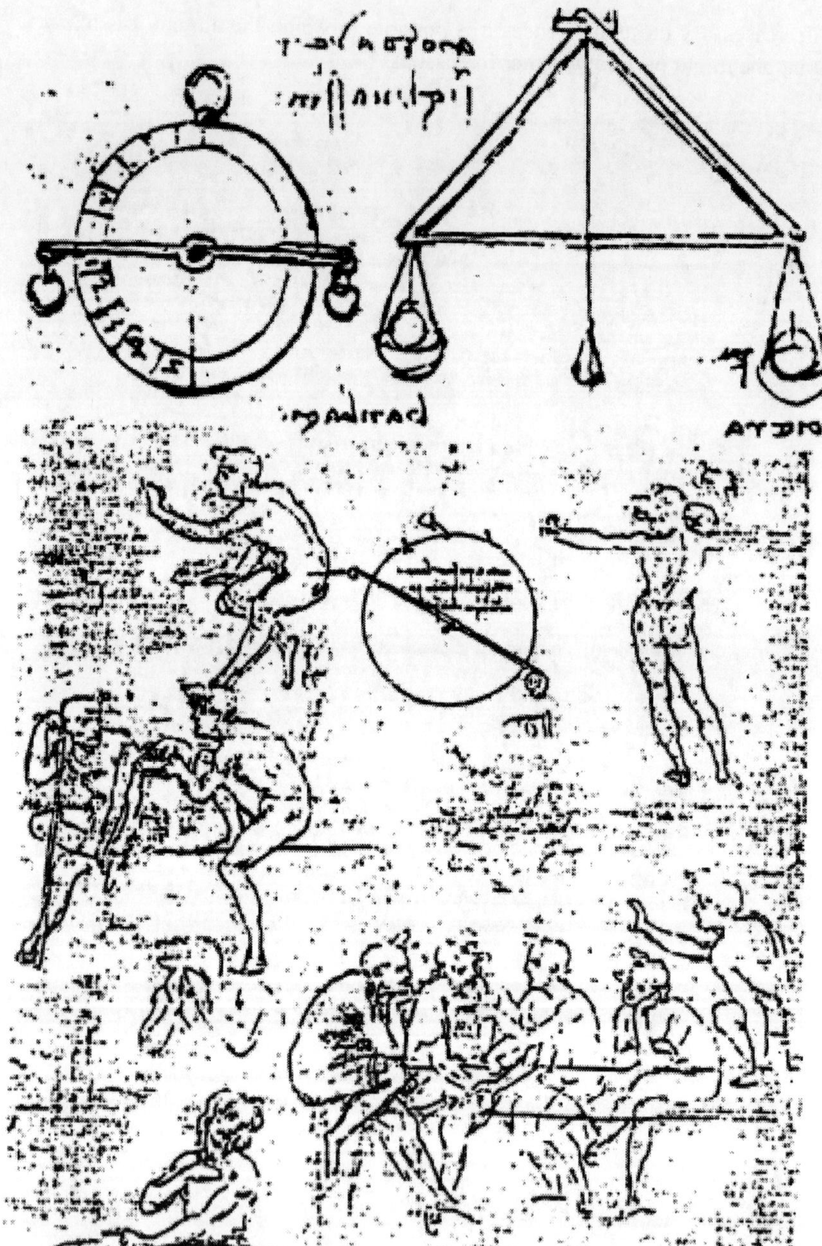

Figure 6. Sketches of gravimetric hygrometers/barometers of Leanardo da Vinci.

About 20 years later the Italian architect and painter Leo Battista Alberti (1404-1472) described a similar device,

> "We know, that a sponge becomes wet from atmospheric humidity and by this fact we make a balance with which we weigh the weight of the air and the dryness of the winds." [14].

From Leonardo da Vinci (1452-1519) we have three designs of inclination balances loaded with a sponge or with cotton (Figure 6) [15-17]. With reflected face he added to the sketches,

> "To recognize the quality and density of the air and to forecast rain."

and

> "Means to detect, when the weather will break-up."

In the following decades a large variety of hygrometers had been developed [1-4].

Definitions

Air humidity is the amount of water vapour in the air disregarding the possible presence of liquid or solid aggregates like raindrops, fog, snow or hailstones. The parameter "air humidity“, which is a function of temperature, is governed only by the availability of water and the thermal energy available for evaporation. Air humidity can be expressed in several ways [18]:

- Absolute humidity AH_{air} is defined as the mass of water vapour m_w presented in unit mass m_{air} or unit volume V_{air} of dry air.

$$AH_{air} = \frac{m_w}{m_{air}} \quad / \text{g kg}^{-1} \tag{1a}$$

or

$$AH_{air} = \frac{m_w}{V_{air}} \quad / \text{g m}^{-3} \tag{1b}$$

- Relative humidity RH_{air} is the ratio of the amount of water vapour in the air m_w at a specific temperature to the maximum mass that the air could hold at this temperature $m_{w,max}$ usually expressed as a percentage.

$$RH_{air} = \frac{m_w}{m_{w,max}} \times 100 \quad /\% \tag{2a}$$

- Another definition is the ratio of the partial pressure of water vapour p in a gaseous mixture of air to the saturation vapour pressure p_0 at a specific temperature. The values differ slightly near the saturation point.

$$RH_{air} = \frac{p}{p_0} \times 100 \quad /\% \tag{2b}$$

- Relative humidity is the most frequently encountered measurement of humidity because it is regularly used in weather forecasts.
- Specific humidity SH_{air} is the ratio of water vapour to air (dry air plus water vapour) in a particular volume of air. Specific humidity ratio is expressed as a ratio of kilograms of water vapour, m_w, per kilogram of air, $m_{air} + m_w$.

$$SH_{air} = \frac{m_w}{m_{air} + m_w} \tag{3}$$

- The dew point is associated with the relative humidity. This is the temperature at which water vapour saturates from gas into liquid or solid usually forming rain, snow, frost, or dew. At this temperature the relative humidity is 100 %. From the dew point temperature, the relative humidity can be derived by means of tables.

Measuring Methods of Air Humidity

The peculiar characteristic of the water molecule facilitates its identification and allows the application of very different methods to determine humidity. The methods are standardised in all industrial states [19].

Water vapour is as a primary contaminant in process gases at very low-levels of particular interest to the semiconductor industry. Therefore, efforts are directed toward developing water vapour measurements and standards at the part per billion level and below for characterisation of moisture concentration. Measuring instruments and standards are normally used in industrial applications for contamination control in process gases.

Table 2. Methods for the Measurement of Air and Gas Humidity

Method	Principle of measurement
	Measuring method
Hygrometer	Measurement of parameters, like change of length, mass, electric resistance, electric capacity, sorption, dew point, that depends on humidity
Psychrometer	Measurement of evaporation coldness by temperature difference
Dew point meter	Observation of condensation at a surface (chilled mirror) in dependence of temperature
Gravimetric train	Elaborate absorption system maintained as the primary standard for humidity measurements
	Humidity sensor
Dielectric measurement	Capacitive measurement with a condenser taking the advantage of the high dielectric constant of water
Electric conductivity measurements	Electrochemical measurement of conductivity
Microwave and infrared spectroscopy	Measurement of absorption of radiation
Nuclear magnetic resonance spectroscopy	Measurement of resonance between a high-frequency electromagnetic field and ^{1}H nucleus of water of a sample, which is arranged into a strong homogeneous magnetic field
Activation analysis	Measurement of absorption of fast neutrons or γ-rays
Moisture indicator	Qualitative test observing colour change

Hygrometry

The earliest hygrometers made in the 15th century were balances loaded on one side with textile [17, 20]. Today the gravimetric method is applied as a primary standard for humidity measurements. The NIST gravimetric hygrometer has an uncertainty of 0.1 % mass fraction. Furthermore, a suitable mass sensor is the quartz crystal balance whose surface is covered by a hygroscopic film or by a microporous substance [21-22]. As the mass of the crystal changes due to adsorption of water vapour, the frequency of the oscillator changes. Ultrathin Linde Type A (LTA- type) molecular sieves grown on a quartz crystal microbalance have high sensitivity, good reversibility and long life at low humidity ranges [23].

The simplest and still widely used mechanical hygrometer is the hair hygrometer of Horace Bénédict de Saussure [24] (Figure 7). It measures the elongation of hair with increasing humidity due to swelling of the hair cells. Hairs from humans, sheep or horses are used; today also plastic strips or fibres are used. Whereas the mass adsorbed is somewhat proportional to relative air humidity the elongation of the hair or fibre (up to ~ 2.5 %) is not linear and the scale is spread in the middle of the measuring range. The traditional folk art device known as a "weather house" works on this principle with a sensor being catgut or hair.

More accurate measurements and electronic recording of results are achieved with a coulometric hygrometer. It measures the change in the electrical impedance or resistance of a thin layer of a hygroscopic material, lithium chloride or phosphorus pentoxide e.g. (Figure 8).

Other hygrometers sense changes in electric capacity (with an electret e.g.), volume, or the transparency of various substances that respond to relative humidity. The oxide sensor is made up of an inert substrate material and two dielectric layers, one of them is sensitive to humidity. The moisture molecules pass through the pores of the surface and cause a change to a physical property of the layer beneath it. An aluminium oxide or silicon oxide sensor has two metal layers that form the electrodes of a capacitor. The number of water molecules adsorbed will cause a change capacity of this condenser due to the high dielectric constant of water.

The measurement of the electrical resistance of thin hygroscopic films or of semiconductors can be used in hygrometry and, likewise, as sensors in head space analysis in order to determine material humidity. Typical substances include lithium chloride, stearic acid and zinc stearate [25]. The relation between conductivity and relative humidity is in general not linear and may be applied only within a restricted measuring range. Highly sensitive humidity sensors based on single SnO_2 nanowires have linear response in conductivity to humid air [26]. Cells with MnO_2 clay solid electrolytes have good linearity and fast response in galvanic potential to changes in humidity [27].

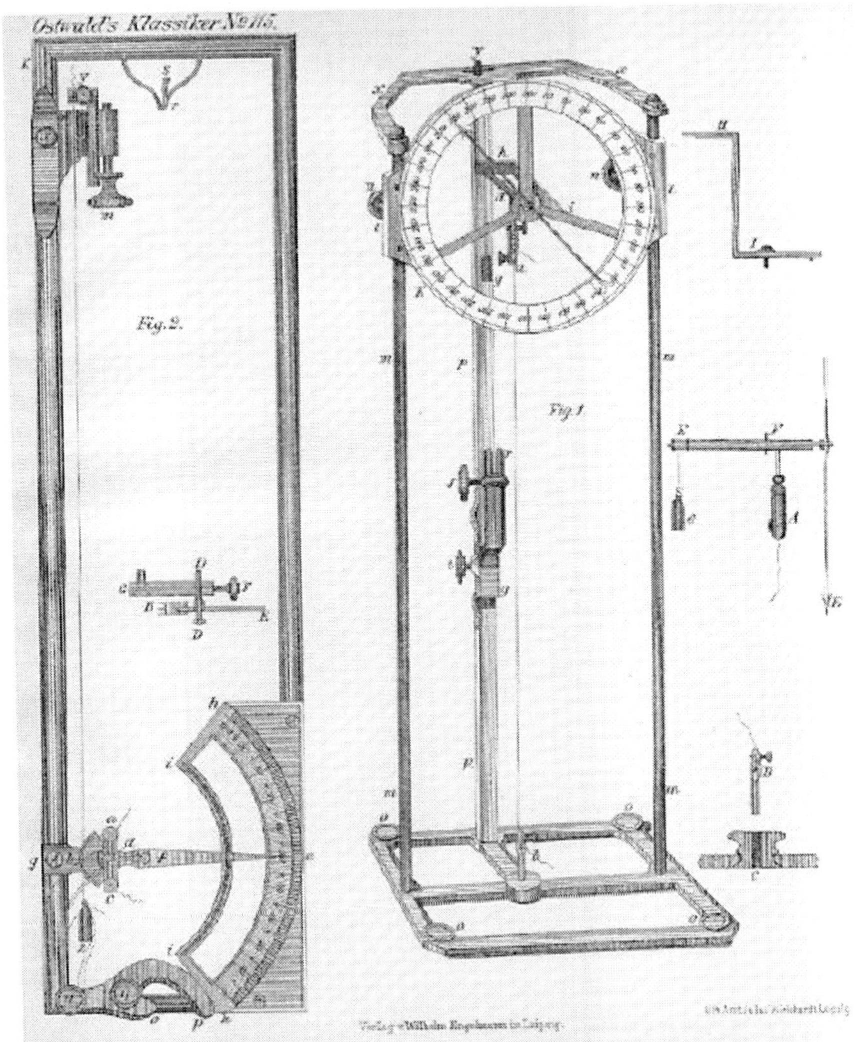

Figure 7. Hair hygrometer of Horace Bénédict de Saussure (1783). Figure 7a: Large hygrometer. The hair is stretched between a and b. Figure7b: Hand hygrometer. The hair is stretched between y and z.

Thick film and disc humidity sensors based on semiconducting hubnerite ($SnWO_4$) compositions show a maximum sensitivity factor when doped with Li^+ andW^{6+} sites in hubnerite material that contributes to hubnerite sensing mechanism [28].

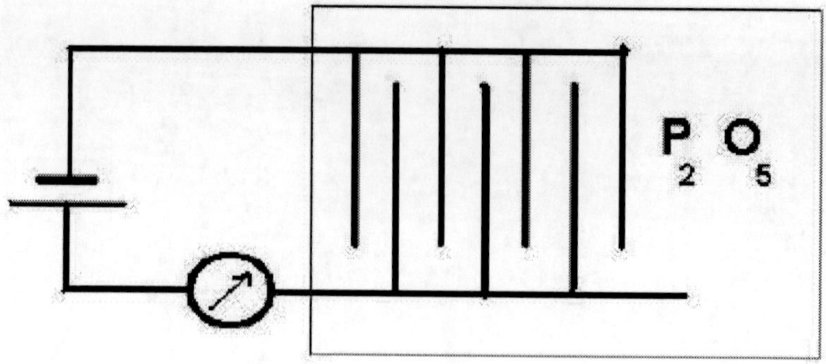

Figure 8. Coulometric hygrometer: Keidel cell (1950): platinum electrodes on a substrate covered with a thin film of P_2O_5.

A silicon oxide sensor can be an optical device that changes its refractive index when water is absorbed into the sensitive layer. A wavelength shift can also be detected on the output, which can be precisely correlated to the moisture concentration.

Optical methods include the observation of colour transition of chemical compounds when reacting with water molecules. They are widely used to control the activity of drying agents and as breakthrough indicators [29]. Test methods of qualitative water indication include redden of blue cobalt(II) chloride [26, 30]. Usually, blue cobalt(II) chloride is dispersed in silica gel. Sensors, which consist of such salts, are dispersed on different substrates, such as cellulose, cellulose acetate, polyvinyl pyrollidon, wool, calcium sulphate, silica gel, zeolites and alumina. Wool impregnated with cobalt chloride is blue in dry state and converts to yellow by moisture adsorption [31]. Humidity indicating gels, which are light blue in dry state and dark blue in moist state, are obtained by dispersing copper sulphate in silica gel. Yellow lead iodide is obtained from a mixture of potassium iodide and lead nitrate by moisture adsorption. Polymer dye systems changing colour with air humidity can also be used as humidity sensors. A high sensitive optical humidity probe has been developed by use of violet Nafion crystal violet films. At a wavelength of 650 nm, a reversible change in the absorbance of the films with relative humidity is linear in the range of zero to 1 % of moisture [32]. For rough control of air humidity in closed rooms, cheap colour indicators on plates are offered.

The colour indicator tube (Dräger Tube or Stain Tube) is used for a quick and rough measurement of moisture in natural gas pipelines. Each tube contains a chemical that changes its colour when the gas passes through. For use, the tube is inserted into a gas pump, used once and discarded. The error can be as high as 25 percent.

In absorption spectroscopy, light passes through a gaseous sample and the amount of light absorbed at a specific wavelength is measured. A tunable laser provides a source of narrow, variable wavelength light that can be used to analyze the small spectral features. According to the Lambert-Beer law, the amount of light absorbed by the gas is proportional to the amount of gas present in the light's path; therefore, this technique is a direct measurement of moisture.

Lyman-alpha hygrometer [33] is a hygrometer based on the absorption of radiation by water vapour at the Lyman-alpha line, which is an emission line of atomic hydrogen at 121.567 nm. Lyman-alpha radiation can be generated by a glow discharge in hydrogen, and a nitric oxide ion chamber normally accomplishes detection. Two magnesium fluoride windows both at the radiation source and at the detector bound the absorption path. Lyman-alpha hygrometers are used on aircraft and on meteorological towers for high-frequency humidity measurements. Inconveniences of the method, like drift of the source intensity or contamination of the windows, are overcome by special calibration techniques or by baselining the high-frequency output to the humidity values provided by a slower, but stable, hygrometer.

Psychrometry

In 1825 August constructed the first psychrometer (Figure 9) on a proposal of Leslie [34-35]. Richard Assmann (1845-1918) added a ventilator, and this aspiration psychrometer [36] is regarded as a standard instrument for air humidity measurements (Figure 10). It consists of two thermometers; the bulb of one of it is kept wet usually by means of a humidified cotton envelope. Evaporation from the wet bulb lowers the temperature, so that the wet-bulb themometer shows a lower temperature T_{wet} than those of the dry-bulb thermometer T_{dry}. Evaporation is supported by an air stream, which is generated either by a ventilator or by whirling the thermometer in the air by hand (sling psychrometer). From the temperature difference, the partial pressure of water vapour in the air $p_{\text{H}_2\text{0}}$ can be determined using Sprung's formula [37].

$$p_{\text{H}_2\text{0}} = p_0 - C_{psy}\,(T_{dry} - T_{wet}) \qquad (4)$$

where p_0 stands for the saturation pressure at the temperature of the wet bulb T_{wet} and the psychrometric constant C_{psy} = 0,67 hPa K^{-1} stands for a height up to 500 m and T_{wet} > 0 °C. The relative humidity can be derived by means of graphical tables.

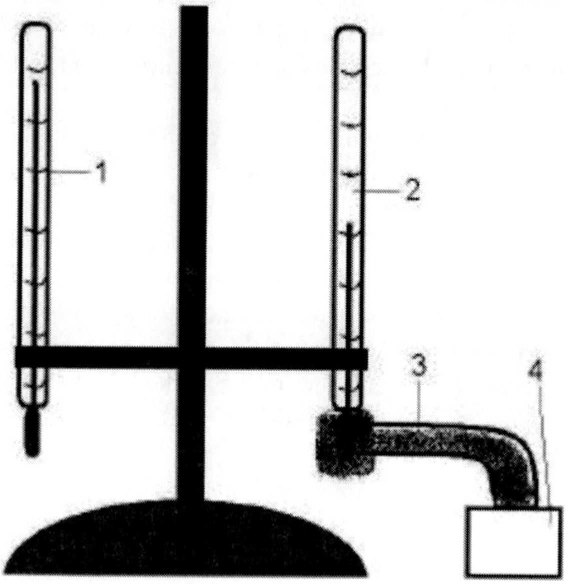

Figure 9. August psychrometer (1825). 1 dry thermometer, 2 wet thermometer, 3 cotton strip, 4 water reservoir.

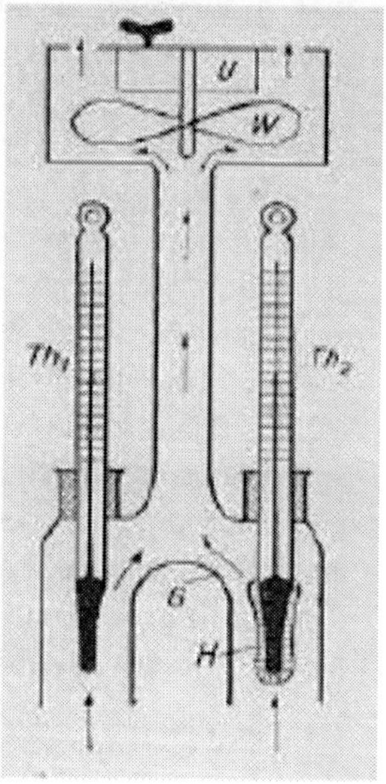

Figure 10. Friedrich Aßmann's aspiration psychrometer (1887). H cotton envelope, G air channel, Th thermometer, W ventilator, U motor.

Dew Point Hygrometry

In 1820, John Frederic Daniell invented a dew point hygrometer, which came into widespread use. The Lambrecht dew point hygrometer is a metal mirror with good thermal conductivity, such as silver or copper. The mirror is properly plated with an inert metal, such as iridium, nickel or gold, to prevent tarnishing and oxidation and is polished. It is cooled by evaporation of ether until moisture just begins to condense onto it (chilled mirror, CMH) (Figure 11). The decrease of light intensity due to scattering is observed optically. The temperature at the beginning of opaqueness of the metal is the dew point temperature.

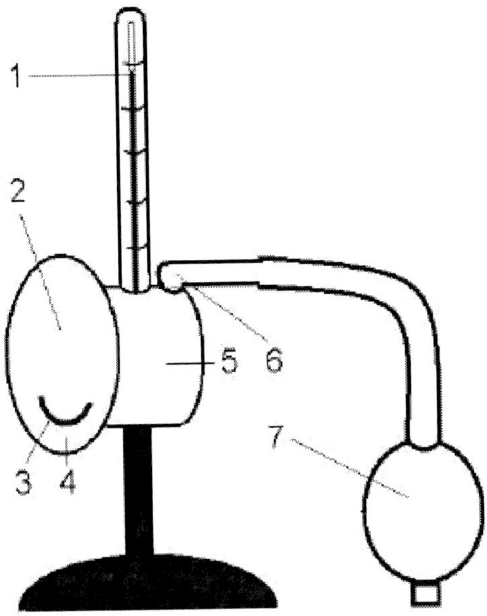

Figure 11. Lambrecht dew point hygrometer (1881). 1 thermometer, 2 metal mirror, 3 gap, 4 reference mirror, 5 container filled with ether, 6 air supply, 7 air pump.

Today the mirror is cooled using a Peltier cooler. The temperature of the mirror is controlled by electronic feedback to maintain a dynamic equilibrium between evaporation and condensation on the mirror, thus, to measure the dew point temperature closely. A beam of light, typically from a solid-state LED, is aimed at the mirror surface and a photodetector monitors the reflected light. A platinum resistance thermometer (PRT), which is properly embedded in the mirror, monitors the mirror temperature at the established dew point. Knowing the atmospheric temperature and the dew-point temperature, the relative humidity can be determined by means of a table. Among the various hygrometric techniques, the dew point method is considered as the most accurate (relative accuracy is about 1 % above 5% RH). Only the gravimetric train, which is used as a primary standard, exceeds it.

In order to increase the accuracy of this type of hygrometer, surface acoustic wave (SAW) devices are very useful because of their dual ability to detect the dew deposition and to measure the temperature with great accuracy [38]. Rayleigh wave properties were investigated in order to study the influence of dew deposition. The experimental device is cooled by means of a Peltier element until dew deposition. Because of polarization of the Rayleigh wave, the dew position induces a substantial attenuation of the wave amplitude and a shift in the wave velocity (mass-loading effect). In order to combine the influence of the temperature and the dew deposition, a thermocouple is deposited on an LST-cut quartz plate. The frequency versus temperature response is used for the dew point detection.

The application of paint and other coatings is very sensitive to humidity and dew point. Condensed water can disturb the operation and can even damage the electronic and micromechanical devices of camcorders and similar apparatus e.g., which are therefore equipped with a dew warning (dew check). Usually, this is an electric sensor whose resistance increases when moisture is present.

Quartz crystal resonator (QCM) dew/frost point sensors operating in the range -90 °C to 15 °C have a fast response time, high sensitivity and accuracy. It was possible to distinguish between supercooled dew and frost from a single scan of quartz crystal resonator [39-40].

Radiometry and LIDAR Hygrometry

LIDAR (light detection and ranging) is an optical remote sensing technology that measures properties of scattered light to find information of a distant target [41-42]. By means of laser pulses and by measuring the time delay between transmission of a pulse and detection of the reflected signal, the distance to an object can be determined. Whereas radar uses radio waves, LIDAR uses much shorter wavelengths of the electromagnetic spectrum, typically, in the ultraviolet, visible, or near infrared region. In general, it is possible to image a feature or object only about the same size as or larger than the wavelength. Thus, LIDAR is highly sensitive to aerosols and cloud particles. An object has to produce a dielectric discontinuity in order to reflect the transmitted wave. Different types of scattering are used for different LIDAR applications, most common are Rayleigh scattering, Mie scattering and Raman scattering as well as fluorescence. The wavelengths are ideal for measurements of smoke and other airborne particles (aerosols), clouds, and air molecules. Since the deployment of the GPS (Global Positioning System) in the 1980s precision positioning has become possible. Airborne LIDAR systems monitor glaciers and have the ability to detect subtle amounts or growth or decline. In atmospheric physics, LIDAR is used as a remote detection instrument to measure densities and movement of certain constituents of the middle and upper atmosphere, such as clouds. Inelastic scattering is observed with Raman spectroscopy. Since the energy change in this case is quantified, the spectrum is characteristic for the target molecule. Water molecules scatter predominantly green light, Nd:YAG laser light with a wavelength of 532 nm is reflected with 660 nm wavelength. This process is applied to determine the relation of a water vapour mixture in the atmosphere.

Radiometers operating in the millimetre or submilimetre range of wavelength have an enhanced sensitivity to low water vapour and liquid contents in comparison to conventional microwave radiometers operating below 30 GZ (1 cm). The temperature and humidity profile up to 5 km above ground is determined in the arctic by using ground-based millimetre wave radiometry and variational retrieval technique [43].

4. Humidity of Liquids

In liquids, traces of water often deteriorate electrically insulating properties or cause corrosion. Humidity can be determined by means of some types of hygrometers described above. For example, an electrical resistance hygrometer with aluminium an aluminium oxide moisture sensor can express the absolute humidity of gases and liquids in dew point temperature, parts per million by volume (PPMv) or parts per million by mass (PPMw). Any gas hygrometer can be used in head space gas analysis or after thermal evaporation. Likewise, some moisture measuring methods designed for solid materials can be applied, in particular thermogravimetry.

5. Moisture of Solid Materials

Historical

The earliest report on a dehydration process is found in Vitruvius (about 70 to 10 BC) "*Books on Architecture*" [44-45]. He describes conscientiously the process of calcination of limestone,

> "The reason why lime makes a solid structure on being combined with water and sand seems to be this: Those rocks, like all other bodies, are composed of the four elements. Those which contain a larger proportion of

air, are soft; of water, are tough from the moisture; earth, hard; and of fire, more brittle. Therefore, if limestone, without being burned, is merely pounded up small and then mixed with sand and so put into the work, the mass does not solidify nor can it hold together. But if the stone is first thrown into the kiln, it loses its former property of solidity by exposure to the great heat of the fire, and so with its strength burned out and set free, and only a residuum of heat being left lying in it, if the stone is then immersed in water, the moisture, before the water can feel the influence of the fire, makes its ways into the open pores; then the stone begins to get hot, and finally, after it cools off, the heat is rejected from the body of the lime.

Consequently, limestone when taken out of the kiln cannot be as heavy as when it was thrown in, but on being weighed, though its bulk remains the same as before, it is found to have lost about a third of its weight owing to the boiling out of the water. Therefore, its pores being thus opened and its texture rendered loose, it mixes readily with sand, and hence the two materials cohere as they dry, unite with the rubble, and make a solid structure."

Thermogravimetry begun with Talabot, who equipped a laboratory with 39 balances for humidity control of silk, which was imported from China by ship, at Lyon in 1833 (Figure 12) [46-47]. The whole equipment including the lovely enamel-work, which shields the ovens, seems to be lost. In 1915 Honda appears to be the very first to use the expression "thermobalance" for his instrument (Figure 13) [48-49]. Soon afterwards such instruments were used to investigate the metabolism of plants (Figure 14).

In 1877 Hannay [50] described a gravimetric method to determine dehydration isotherms of salts. He conveyed water vapour, which was released from a heated sample into a cylinder that was filled with an ad- or absorbent, by means of a carrier gas.

Figure 12. Désiccateurs Talabot 1833. Conditioning apparatus at Lyon for the measurement of the humidity of silk.

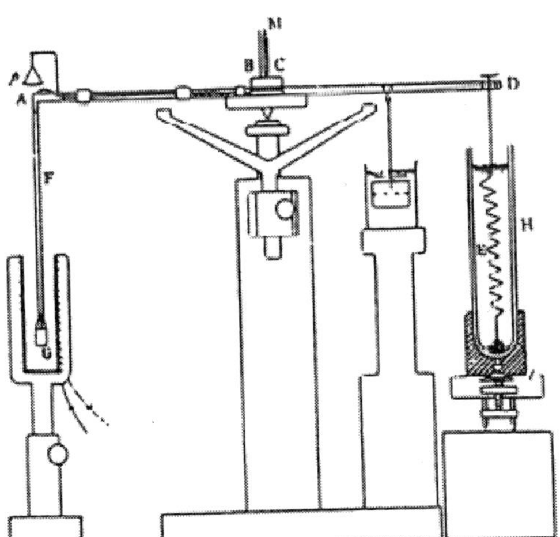

Figure 13. Honda's thermobalance.

Figure 14. Gravimetric instruments for the investigation of the metabolism of plants.

The cylinder was weighed periodically. Warming up samples in a heating chamber and sequential measurements of the decreasing sample mass in order to determine the desorption isotherm is still widely used [2-3].

In 1886 Warburg und Ihmori [51] built a beam microbalance for adsorption measurements. The first vacuum microbalances with electromagnetic compensation were made by Emich [52-53] and Urbain [54] in 1912. Today a variety of electrodynamic compensating vacuum microbalances is available on the market [55-56]. Several apparatus equipped with such balances for automatic measurement of adsorption isotherms are offered also. The first one has been built in 1962 [57] (Figure 15). Many of those microbalances were used first to measure water adsorption from the atmosphere. Today the measurement of humidity is facilitated by so called drying or moisture balances, which are laboratory scales equipped with an infrared or microwave heater (Figure 16) [7].

Material Moisture Measuring Methods

There are indeed two types of humidity: Air humidity, which is independent from the other components of the air, and the humidity of materials, which is influenced or even dependent on the material in question that may be a solid, a liquid or a gas.

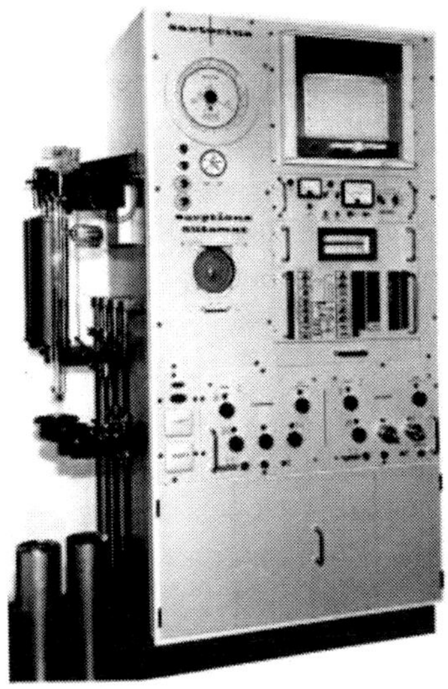

Figure 15. Sartorius Sorptomat of Sandstede and Robens. Apparatus for the automatic measurement of adsorption isotherms. On the left a Gast vacuum microbalance sticks.

Figure 16. Drying or moisture balance Sartorius, with halogen infrared heater.

The common term humidity also comprises liquid contents other than water, solvents e.g. [58]. In the following, we restrict ourselves to the content of water (moisture) generally expressed as the amount of water in relation to unit mass or unit volume of the sample.

Table 3. Methods for the measurement of moisture content

Method	Principle of measurement
Oven-drying Moisture balance	Thermal activation of the sample and gravimetric measurement of mass loss
Thermogravimetry	Mass loss under controlled temperature programme
Sorption measurement Desiccator method	Variation of (partial) water vapour pressure and measurement of mass change
Standard contact porometry	Contact with standard of defined humidity, measurement of mass change
Dielectric measurement	Capacitive measurement with a condenser taking the advantage of the high dielectric constant of water
Electric conductivity measurements	Electrochemical measurement of conductivity
Microwave and infrared spectroscopy	Measurement of absorption of radiation
Nuclear magnetic resonance spectroscopy	Measurement of resonance between a high-frequency electromagnetic field and ^{1}H nucleus of water of a sample, which is arranged into a strong homogeneous magnetic field
Activation analysis	Measurement of absorption of fast neutrons or γ-rays
Calorimetric method	Measurement of heat capacity and conversion heat
Karl Fischer	Titration using Karl Fischer reagent
Phosphorus pentoxide	Thermal activation of the sample, absorption of water
Moisture indicator	Qualitative test observing colour change

The peculiar characteristic of the water molecule and its anomalous features facilitates its identification and allows the application of very different methods to determine water content and humidity depending on the ambient conditions. However, only a few methods provide reliable values [59] of the water content of solids and enable to distinguish between free bulk water, adsorbed vicinal water, absorbed and chemisorbed water. Furthermore, just the gravimetric methods measure the sum of volatile substances within a solid, whereas, in general, the task is to determine the water content only. Mostly applied methods used as routine tests are summarised in Table 3.

Drying Methods

The classic laboratory method of a moisture measurement of solids is the determination of the loss on drying (LOD). On the other hand, drying of the sample is necessary in order to adjust definite starting conditions for adsorption experiments. In both cases careful methods, which avoid damage of the sample, should be applied. The following methods have proved to be useful.

Drying at Elevated Temperature

A temperature as high as possible at which the sample is not damaged is determined by pilot tests using thermogravimetry. The drying process can be controlled thermogravimetrically as well.

Vacuum Drying

Vacuum is applied at a pressure below saturation pressure of water at ambient or elevated temperature. The degree of dryness may be assessed by observing the pressure increase after locking of the connection valve to the vacuum pump.

Drying by Means of Desiccants

The sample is placed in a desiccator equipped with a desiccant, usually silica gel with moisture indicator, phosphorous pentoxide (p-drying), magnesium perchlorate or concentrated sulphuric acid.

Cold Trap Drying

At ambient temperature, the sample vessel is connected to a cold trap, which is cooled by means of a mixture of dry ice with alcohol (d-drying) or a refrigerator. In addition a vacuum pump may be connected to the cold trap.

Gas Flow Drying

An inert gas is passed above the sample usually at elevated temperature. The procedure is time consuming.

Freeze Drying, Lyophilization

Usually, the material is placed in a freeze-drying flask that rotates in a bath called a shell freezer, which is cooled below the triple point of water by mechanical refrigeration, a dry ice and methanol mixture (193 K) or liquid nitrogen (77 K). Then the frozen water in the material sublimes directly from the solid phase to the gas phase by reducing the surrounding pressure with a vacuum pump and adding enough heat. To avoid recrystallisation, sensible materials are frozen very fast.

Solvent Exchange

The water included in the material is replaced stepwise by other solvents, methanol, ethanol, isopropanol, alcohol solutions in increasing concentrations 30 %, 50 %, 60 %, 70 %, 80 %, 90 % 96 % DMSO or diethlyether e.g. The final solvent is vaporised at elevated temperature. The last solvent should be volatile and should have a low surface tension. This complicated and slow method avoids damage by crystallization particularly of biological materials.

Supercritical drying is a process to remove liquid in a precisely controlled and gentle way. It is useful for the production of aerogels and for the preparation of biological materials. When a substance crosses the boundary from liquid to gas the substance volatilises and so the volume of the liquid decreases. Sensible structures like cell walls collapse because of the surface tension at the solid-liquid interface. To avoid this, the sample can be brought from the liquid phase to the gas phase without crossing the liquid-gas boundary on the phase diagram by passing through the supercritical region at high temperature and pressure above the critical point. Densities of the liquid phase and vapour phase become equal at the critical point of drying. Fluids suitable for supercritical drying include carbon dioxide (critical point 304.25 K at 7.39 MPa) and freon (about 300 K at 3.5-4 MPa). Nitrous oxide has a similar physical behaviour to carbon dioxide, but is a powerful oxidizer in its supercritical state. Supercritical water is also a powerful oxidizer partly because its critical point occurs at a very high temperature and pressure (647 K and 22.064 MPa).

Gravimetric Measurement of Material Moisture

Gravimetric measurements to determine the humidity or dry mass of a solid material are based on the removal of water by reducing the partial pressure of water vapour of the gaseous phase above the sample. This may be done with a vacuum pump, by a condensation process within a desiccator or by means of a dry gas flow. Controlled heating of the sample shortens the measuring time. In addition to physisorbed water, chemically bound water may also be removed. The mass decrease of the solid sample is measured gravimetrically or the water mass removed is weighed.

Gravitational balances measure not the mass but the sum of mass plus buoyancy of the sample. The buoyancy of the adsorbate must also be taken into account in case water is predominant in spongy structures in comparison to the skeleton. Measurements are disturbed by convection and, at low pressures, by thermal gas flow. During the drying of large samples shrinkage of porous materials or agglomeration of fine materials result in encrustation. This obstructs evaporation from interior and distorts the drying process.

Oven-drying is a widely used method. Here the sample is dried at a constant temperature. Humidity is removed by circulating air. The sample is weighed after reaching mass constancy. Such measurements give reliable "true" results of the moisture content only in case of well known drying characteristics of the material. Weighing at intermediate times and deriving a kinetic curve may obtain more information. However, very different results may also be obtained by this method due to the different binding types of water (Figure 17).

Because the manual laboratory method is relativcly slow, automated moisture analysers that can reduce the time necessary for a test from a couple hours to just a few minutes have been developed. Electronic moisture balances are equipped with an infrared or microwave heater and a device to remove the moistened air. The moisture balance (Figure 16) is a simplified model of a thermogravimetric apparatus, also known as conditioning apparatus.

A thermogravimetric apparatus (Figure 18) consists of a balance and a heating unit to adjust the sample temperature at a constant value or to control the defined temperature increase [60]. Measurements are made either in air, an inert gas flow or vacuum [61].

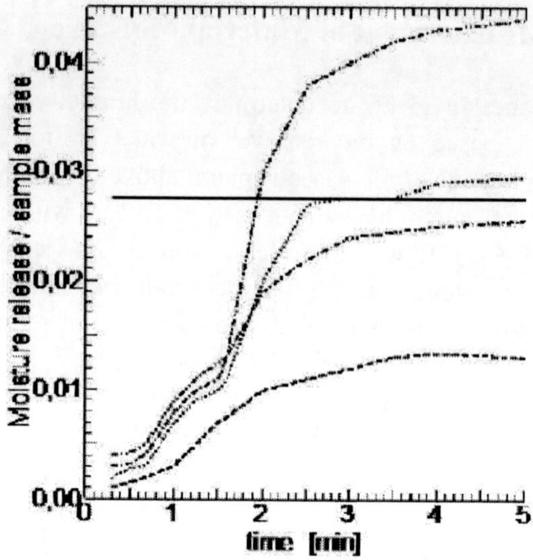

Figure 17. Infrared drying of candies at ---- 90°C, -·-·- 95°C, ····· 100°C, -··-··-·· 110°C, and water content measured using the Karl Fischer method ———, © Isengard.

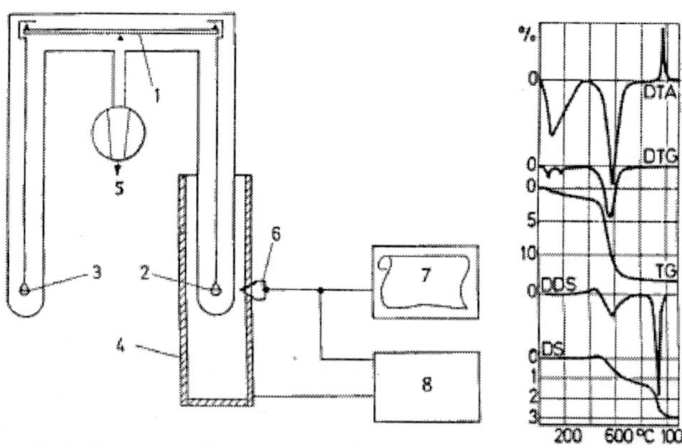

Figure 18. Diagrammatic view of a thermogravimetric apparatus with vacuum thermobalance (MOM Derivatograph). 1 balance beam, 2 sample, 3 counterweight, 4 oven, 5 vacuum pump, 6 temperature sensor, 7 recorder, 8 temperature program control. The records for thermogravimetric analysis (TGA): DTA differential thermal analysis, DTG differential thermogravimetry, TG thermogravimetry, DDT differential thermal dilatometry, thermal dilatometry.

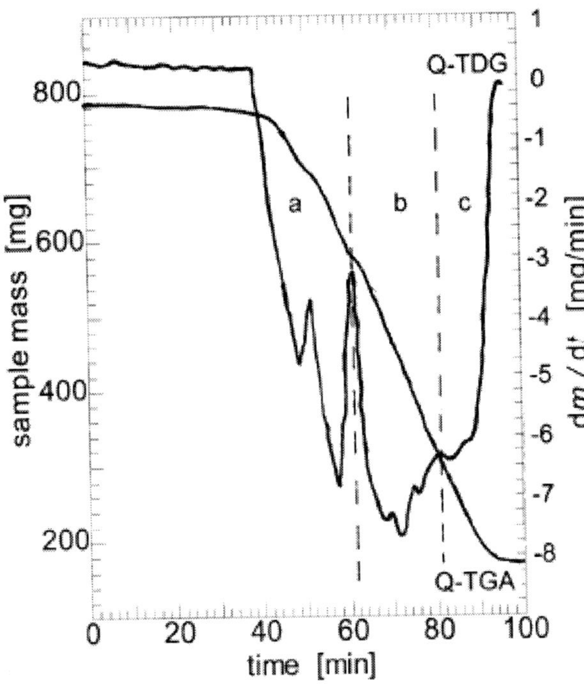

Figure 19. Quasi-isothermal water desorption from activated carbon (Merck) according to Staszczuk. Mass loss (Q-TGA) and its time derivative (Q-DTG) are plotted. © Stasczuck.

To investigate the water content, measurements are performed at low temperatures. Physisorbed water and condensed pore water vaporise mainly up to 100 °C. Chemisorbed components and crystal water are liberated at higher temperatures. Quasi isothermal thermograms (QTGA) are measured at slow temperature increase whereby the sample is held under saturation vapour pressure using special sample containers (Figure 19). QTGA can be evaluated for binding parameters and for pore size distribution.

Measurement of Water Sorption Isotherms

To measure adsorption isotherms of water vapour, the mass adsorbed is usually determined gravimetrically as a function of stepwise varied pressure at constant temperature [62]. Alternatively, the mass adsorbed may be determined from volume, pressure and temperature in the sample vessel using the equation of state. Adsorption measurements are started from a dry sample state in vacuum or dry atmosphere as well as from a defined humidity. Desorption is started from a defined humidity, from saturation pressure if possible.

Sorption isotherms may be measured simply by placing the samples in a desiccator at constant temperature. Different humidities are adjusted by means of salt solutions [63]. One sample is exposed to a defined single humidity by the integral sorption method whereas the humidity around one sample is stepwise varied by the interval method. Intermediate evacuation or movement of the gas atmosphere can speed up measurements.

An apparatus of gravimetric water sorption consists of a microbalance and a thermostat. Water vapour pressure is adjusted and varied by means of a carrier gas flow loaded with water vapour (Figure 20a). Alternatively, using a vacuum balance, either portions of water vapour are added or the pressure is adjusted by means of a thermostated water reservoir (Figure 20b).

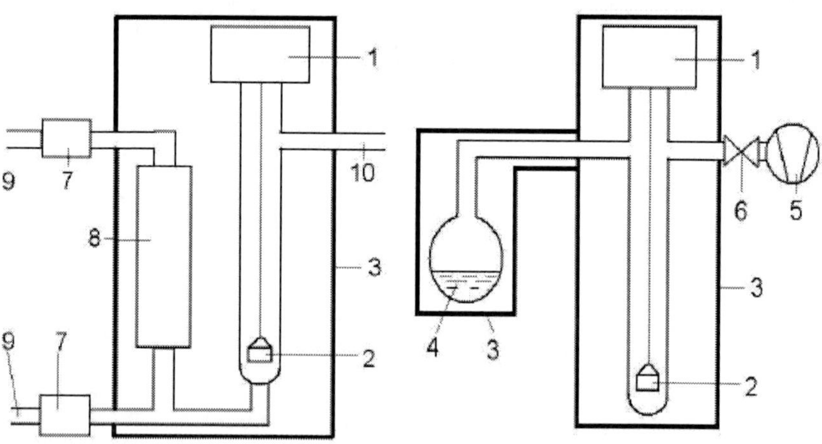

Figure 20. Apparatus of gravimetric water sorption consisting of a microbalance and a thermostat. 1 vacuum balance, 2 sample, 3 thermostated case, 4 water, 5 vacuum pump, 6 valve, 7 flow control, 8 humidifyer. 9 gas inlet, 10 gas outlet.

Figure 20a. Water vapour pressure is adjusted and varied by means of a carrier gas flow loaded with water vapour.

Figure 20b. Alternatively, using a vacuum balance, either portions of water vapour are added or pressure is adjusted by means of a thermostated water reservoir.

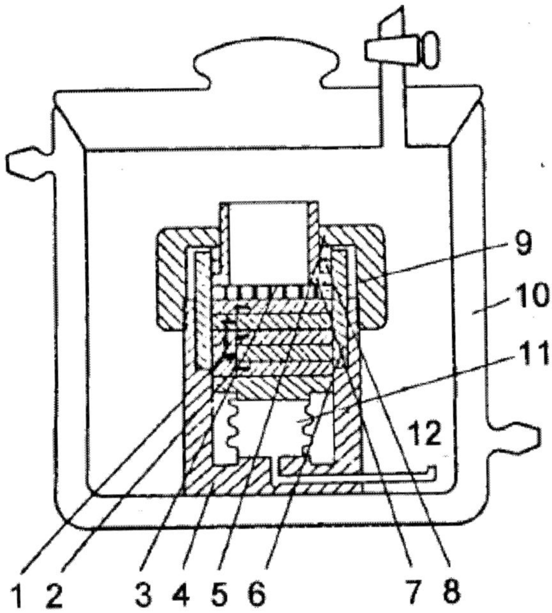

Figure 21. Conditioning equipment for Standard Contact Porometry according to Volfkovich, Bagotzky, Sosenkin and Shkolnikov. 1 sample layer, 2 standard layer, 3 diaphragm, 4 vessel, 5 screw flange, 6 centring tube, 7, 12 vapour inlet, 8, 9 O-rings, 10 thermostat, 11 bellows.

Modified volumetric apparatus as widely used for the measurement of adsorption of permanent gases may also be applied for the measurement of water sorption isotherms. Here, condensation in the manifold must be avoided. Drying processes can be easily observed by stepwise pumping of the water vapour, which is developed from a sample, in a calibrated volume.

The standard contact porometry was developed by Volfkovich, Bagotzky, Sosenkin and Shkolnikov [64]. Here, the sample is brought into contact with a porous standard sample with defined water content (Figure 21). In thermodynamic equilibrium, the liquid into the whole pore system has an identical chemical potential. The humidity of the sample can be derived from the equilibrium water mass in the standard. A set of standards allows the determination of an isotherm even near saturation pressure.

In water adsorption isotherms, the specific adsorbed amount n_a is plotted as a function of the relative vapour pressure p/p_0.

$$n_a = n_a\left(\frac{p}{p_0}\right)_T \tag{5}$$

This corresponds to eq. (2b) applied in hygrometry [18]. The specific adsorbed amount n_a is derived from the gravimetrically or volumetrically measured data using

$$n_\mathrm{a} = \frac{m_\mathrm{a}}{m_\mathrm{s}\cdot M_\mathrm{a}} = \frac{p_\mathrm{n}}{R\cdot T_\mathrm{n}}\cdot V_\mathrm{a} \tag{6}$$

where m_a is the mass adsorbed, m_s the sample mass, M_a is the mole mass of the adsorbate water, p_n is the standard pressure, R is the universal gas constant, T_n is the standard temperature, V_a is the specific adsorbed gas volume.

With hydrophilic surfaces, isotherms of types I, II, IV and VI according to IUPAC classification (Figure 22) [6, 65] are observed. The hysteresis may be voluminous and may cover the whole region of relative pressure [32, 66]. On hydrophobic materials, adsorption starts only at elevated pressures; that means there is hardly adsorption at the surface but condensation in pores and isotherms correspond to type II and V. However, often water sorption isotherms cannot be assigned to any type of the IUPAC classification as demonstrated in Figure 23.

From type I, II and IV isotherms, the specific surface area [67-68] can be calculated by the method of Brunauer, Emmet and Teller (BET) [69] . Type I may be described by the Langmuir equation and the horizontal plateau results from a monolayer covering the surface (monolayer capacity n_m). The monolayer capacity n_m is calculated using the BET equation:

$$\frac{\frac{p}{p_0}}{n_a\left(1-\frac{p}{p_0}\right)} = \frac{1}{n_m C} + \frac{C-1}{n_m C}\cdot\frac{p}{p_0} \tag{7}$$

The specific surface area a_s is calculated using

$$a_\mathrm{s} = n_\mathrm{m}\, a_\mathrm{m}\, N_\mathrm{A} \tag{8}$$

where N_A is Avogadro's number and a_m is the molecular cross sectional area occupied by an adsorbed molecule in a complete monolayer. For water a value of $a_m = 0.125$ nm^2 is recommended.

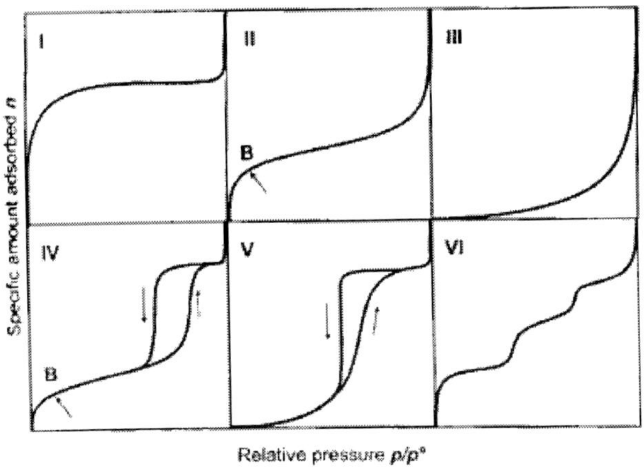

Figure 22. IUPAC classification of gas adsorption isotherms.

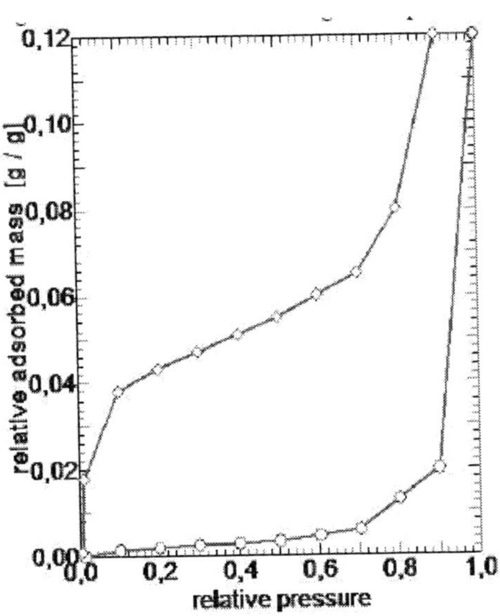

Figure 23. Water vapour isotherm at 298 K on a hardened cement paste (Dyckerhoff white).O adsorption, ◊ desorption.

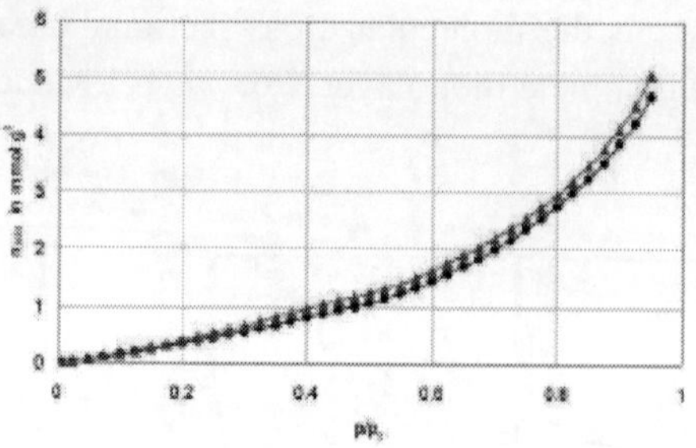

Figure 24a. Water vapour isotherms type III on hydrophobic materials (polyurethane • and silica/polyurethane nanocomposite ▲).

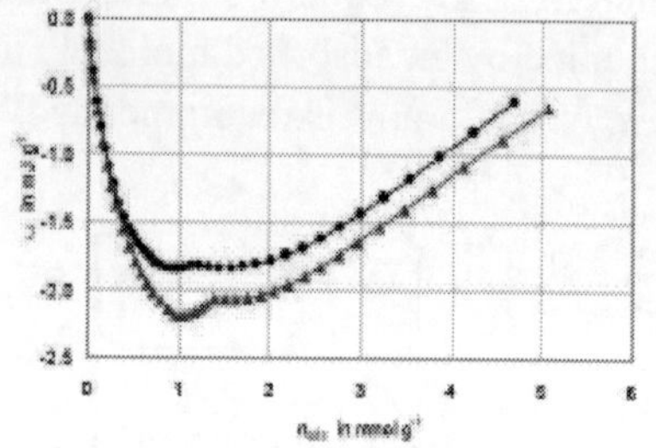

Figure 24b. Transformation of water vapour isotherms in Figure 24a into ESW isotherms.

The minimum of the ESW isotherms corresponds to a loss of degrees of freedom of -1.6 RT/2 for polyurethane • and -1.8 RT/2 for silica/polyurethane nanocomposite ▲. The calculated specific surface areas are 5.9 m^2 g^{-1} (polyurethane) and 6.6 m^2 g^{-1} (nanocomposite), respectively.

The ESW theory by Adolphs [70-72] provides a modeless way of calculating surface energies and specific surfaces areas directly from sorption isotherms. Thermodynamically, the excess surface work (ESW) is the sum of the surface free energy and the isobaric isothermal work of sorption. The measured isotherms are transformed into ESW isotherms by plotting the product of the adsorbed amount and the change in chemical potential versus the adsorbed amount. The ESW isotherms show at least one minimum that yields a sorption energy, which corresponds approximately to the loss of degrees of freedom of the sorptive. Thus, the binding strength of the adsorption layer can be estimated. From the mass adsorbed in the first minimum a specific surface area similar to the BET surface area can be obtained.

Figure 24a shows the water vapour isotherms on polyurethane and silica/polyurethane nanocomposite [73-74]. These are type III isotherms that are characteristic for hydrophobic materials. According to the classical models of BET and BJH, further evaluation of the isotherms is not possible. The transformation into ESW isotherms shows Figure 24b. The minimum of the ESW isotherms corresponds to a loss of degrees of freedom, which describe a medium strong physisorption of the water molecules on the surface. However, the deeper and sharper ESW minimum of the nanocomposite shows that the nanocomposite better adsorbs the water because of the silica additive slightly reduces the hydrophobic characteristics. Further evaluation of ESW isotherm yields a specific surface area of about 6 $m^2 g^{-1}$.

Furthermore from types II and IV isotherms, the pore size distribution can be determined according to the method of Barrett, Joyner and Halenda (BJH) [75] e.g. and various parameters of pore systems can be derived.

Figure 25 shows the water vapour isotherm and the nitrogen isotherm on activated carbon [67]. The whole isotherms as well as the calculated specific surface areas and the pore size distributions are different. However using the adsorbed volume as ordinate, the nitrogen and water sorption curves meet each other at the saturation point in accordance to Gurwitsch's rule [76]. This demonstrates that parameters obtained by using inert gases can only be applied with restrictions to predict adsorption of water. If a material should be used in a humid environment water sorption measurement are unavoidable.

Measuring the adsorbed water mass in a material can give the relative humidity of the air with which the material is in equilibrium. For example, the adsorption isotherm of water vapour on wool in Figure 26 can be used for air humidity measurement at 15 °C and 25 °C [30].

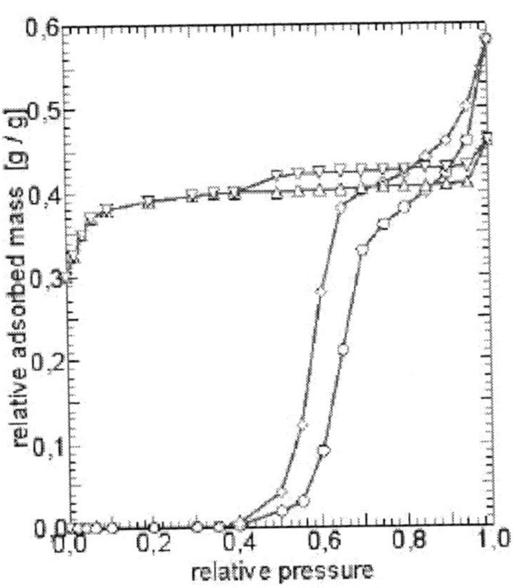

Figure 25. Nitrogen adsorption Δ and desorption ∇ isotherm at 77 K and water vapour adsorption O and desorption ◊ isotherm at 298 K on activated carbon according to Juhola. Note: Using the adsorbed volume as ordinate the nitrogen and water sorption curves meet each other at the saturation point in accordance to Gurwitsch's rule.

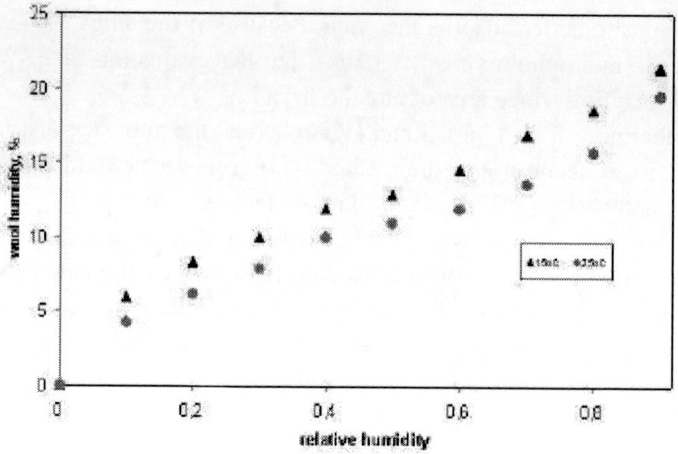

Figure 26. Adsorption isotherm of water vapour on wool at 298 K O and at 288 K Δ.

Calorimetrical Measurements

Using microcalorimeters, the heat capacity of moist and dry samples can be compared and the conversion heat during drying or water adsorption can be observed. Such measurements allow the determination of adsorption isotherms and their evaluation is discussed above.

Thermoporometry [77] makes use of the triple point depression of a liquid in contact with the solid phase within a porous matrix . This effect also depends on the curvature of the interface and, thus, of the pore radius. The sample is immersed in water or benzene and the amount of the frozen liquid is measured as a function of decreasing temperature at ambient pressure by means of a high sensitive calorimeter. From this freezing thermogram, the pore size distribution and the specific surface area can be calculated.

Chemical Methods for Determination of Water Content

Spot analysis for the detection of water include the reddening of blue cobalt(II) chloride or production of yellow lead iodide from a mixture of potassium iodide and lead nitrate (see Chap. "Hygrometry").

The most important method to determine the complete water content is Karl Fischer titration [78-82]. Water is titrated using Karl Fischer's reagent, which consists of iodine, sulphur dioxide, a basic buffer and a solvent. The original composition was modified and adapted to the matter to be investigated. An alcohol, mostly methanol, is usually used as solvent. It is esterified by means of SO_2. To obtain a quantitative reaction the ester is preferbly neutralised by imidazole to yield alkyl sulphite. In a second step, alkyl sulphite is oxidised by iodine to form alkyl sulphate in a reaction that requires water. Simplified the reaction can be written as

$$SO_2 + I_2 + 2\,H_2O \rightarrow H_2SO_4 + 2\,HI \quad (9)$$

The consumption of iodine is measured either coulometrically or volumetrically. In the coulometric procedure, iodine is formed from iodide by anodic oxidation. In the volumetric procedure, an iodine solution is added and different techniques are used for titration. The indication of the endpoint is based on an electrochemical effect in both cases: Two platinum electrodes placed into the working medium are polarised either by a constant current or a constant voltage.

Karl Fischer titration requires water to be in direct contact with the reagent. For insoluble or hardly soluble samples a special treatment is necessary; the water has to be released from the matrix first. Some materials may cause erroneous effects by side reactions.

Because many gases include amounts of water of only few $\mu g\ dm^{-3}$, the admission of the gas sample should be made carefully in order to avoid contamination or adsorption of water by the tubing system [83]. Subsequently, the determination of humidity is made according to ISO 10101 [84-86].

Water vaporised from the sample by controlled increasing temperature can be analysed coulometrically. Parallel arranged electrodes, which are coated with thin layer of phosphorus pentoxide, might be used. Hydrogen ions are neutralised and hydrogen is set free at the cathode whereas the reaction

$$P_4O_{12}^{4-} \rightarrow P_4O_{10} + O_2 + 4e^- \quad (10)$$

takes place at the anode. The current is measured. The resolution of a commercially offered apparatus is 100 ng. The results should be comparable to those of thermogravimetric measurements. In contrast, exclusively the water release is indicated. A typical device is the Keidel cell [7, 87-88], which is a coulometric sensor with platinum electrodes on a substrate covered with a thin film of phosphorus pentoxide [7].

Capacitance Measurements

Electrical capacitance tomography (ECT) is applied for online measurement of solid moisture content of fluidised beds [89]. A twin plane ECT sensor with eight electrodes in each plane was mounted in the bottom of a glass fluidisation chamber. From the adjacent electrode pairs, the water content of the solids was estimated on the basis of the correlation between the moisture content and the permittivity value. To reduce measurement errors, the effect of the temperature on the moisture measurement was compensated.

Measurement of Material Moisture Via the Gas Phase

The material moisture can be measured by evaporation or condensation of sample water and by determination of the vapour pressure and the relative humidity, respectively, within the sample vessel or in a confined volume. This may be performed also stepwise in order to obtain a desorption isotherm. Such a procedure is applied in the phosphorus pentoxide method. Likewise, other air humidity sensors can be used as described in Chapter 3 [7].

Shortening of the Measuring Time

In statically drying or adsorption measurements, the temperature is held constant and the water vapour pressure is changed stepwise to zero or to a given value. Drying and adsorption at constant temperature are generally slow processes. In particular, water adsorption, absorption and drying occur very slowly and the equilibrium or the dry state cannot be reached in a reasonable time. Therefore, often the measuring procedure is shortened without waiting for equilibrium conditions. To speed up measurements and, nevertheless, avoiding errors, an extrapolation method as described by Jäntti in ref. [90-92] should be preferred.

In many cases, the kinetic drying curve follows approximately a simple exponential law:

$$m_a(t) = m_{as}\left(1 - e^{-t/\tau}\right) \tag{11}$$

where $m(t)$ is the sample mass as a function of time t, m_{as} is the asymptotical equilibrium value and τ is the so-called characteristic time of the system solid/water [91, 93]. Establishing the function

$$J(t) = m_a - \frac{(dm_a/dt)^2}{d^2m_a/dt^2} \tag{12}$$

and differentiation reveals a constant value.

$$J(t) = const = m_{as} \tag{13}$$

For its practical application, three values of the initial ad- or desorption curve in equidistant time intervals are measured and $J(t)$ is determined using

$$J(t) = \frac{m_{a2}^2 - m_{a1}m_{a3}}{2m_{a2} - m_{a1} - m_{a3}} \tag{14}$$

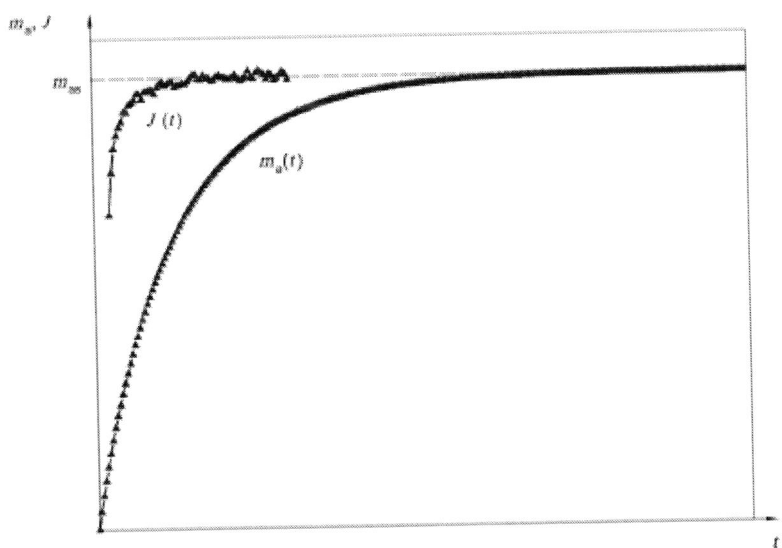

Figure 27. The Jäntti-function $J(t)$ is not constant but approximates the asymptotical equilibrium value much faster then the measured adsorbed mass $m_a(t)$.

Indeed, the experimental Jäntti-function is not constant, but it approximates the asymptotical equilibrium value much faster than the measured adsorbed mass (Figure 27). If the function approximates not quickly a constant value this method is not applicable and it is a definite indication that a complex drying or sorption process occurs, including swelling or encrustation e.g.

6. Extraterrestrial Water

The universe is cold. Celestial bodies, like planets, moons and asteroids, far from their star are usually covered with frozen volatiles, water ice and organics. Dwarf planets Ceres, Haumea and Orcus are covered with water ice. It is believed that comets consist of blocks of refractory material, regolith and frozen volatiles, mostly water [94-95]. Most probably, liquid water can only exist on a celestial body, which is situated in a comparatively narrow zone in vicinity of a star. In that "habitable" zone, a brisk exchange of water takes place between the body and space, well comparable with technical sorption and drying processes discussed in this paper.

In our solar system, today such conditions are found on Earth, Moon, Mars and some other objects, in particular on the Jupiter moon Europa and the Saturn moon Enceladus.

A celestial body was developed from molten material, which initially included no water, but was and still is supplied with water by the following mechanisms:

- The molten basic material contains the components hydrogen and oxygen for the synthesis of water. This process can be observed in volcanic eruptions, in which always water vapour is set free.
- The solidified surface is supplied with water when hit by spatial bodies. Likewise, cosmic dust particles are often covered with ice.
- By its magnetic field, Earth is protected from solar wind, but not the Moon. Solar wind contains mainly electrons and protons. The rocks and regolith that make up the lunar surface consist of about 45 percent oxygen combined with other elements, such as mostly silicate minerals. Protons join with electrons of the surface material to become hydrogen atoms [96]. Some fractions of protons, which are

travelling with one-third the speed of light, have enough force to break apart oxygen bonds in soil materials to form water.

- The opposite process is the evaporation of hydrogen, oxygen and water vapour from the body, which is hindered by gravitation and by atmospheric conditions as the recombination of water molecules to denser aggregates e.g.

Search for extraterrestrial water as well as determination of water inventory and soil moisture is based on remote measurements and analysing the spectrum of electromagnetic radiation either of natural sources or produced by a transmitter and reflected. Spacious deposits have been detected photographically. Robots make in situ investigations and, in addition, soil samples of the Moon are analysed using conventional methods.

Moon

The Moon remains drier than any desert on Earth. Nevertheless, water exists on the Moon:

- Frozen pockets of water ice remnants of water-rich comet impacts are embedded unmelted in the permanently shadowed regions near the poles [97]. Due to the very slight tilt of the Moon's axis, only 1.5°, some of these deep craters never receive any light from the Sun; they are permanently shadowed. Estimations amount to from about 6×10^9 up to 10^{12} kg of water ice in total. At least 0.6×10^{12} kg could be entombed in North Pole craters.
- Volatiles vaporise quickly into the tenuous atmosphere (10^{-10} Pa), influenced by solar radiation and daily surface temperatures up to 130 °C, and vanishes into space due to weak gravity. However, during night the temperature decreases down to -160 °C and some of the imported water is adsorbed. Weakly bound physisorbed water may comprise up to two molecular layers at the surface. Water condensed in pores is bound stronger. Some mineral surfaces hydratise and water is chemisorbed. If such a surface is hidden from solar radiation some of those stronger bound water may be hold for long time. It is estimated (optimistically) that one tonne = 10^3 kg of the top layer of the lunar surface would hold about 1 kg of water.

- Some of the vaporised water may drift to the poles and be stored in craters, which act as cold traps with temperatures down to -230 °C.

First photographic results of the Clementine mission and Lunar Prospector [96] suggested that water is on the Moon. Neutron radiation was observed as a response to the bombardment of cosmic radiation on the surface. Fast neutrons are slowed down by hydrogen and then ice can be detected with a content of 0.1 per mille in the surface material down to a depth of about 0.5 m. The distribution of the neutron radiation gave evidence for the presence of water distributed on the entire surface, but with higher concentrations at the poles. Similar effects are produced by structured surfaces of dry stones and, thus, comparative measurements had been necessary for corrections. The Deep Impact spacecraft made infrared detections of water and hydroxyl groups as part of calibration exercises in 2010.

After termination of its mission, in 1999 the Lunar Explorer and in 2006 the Smart-1 were navigated into such craters, but no water was detected. However, NASA's Lcross sattelite kicked up at least 95 litres of ice and vapour from a South Pole crater.

A NASA radar aboard India's Chandrayaan-I lunar orbiter found 40 craters ranging in size from 1 to 15 km across with pockets of ice. The radar called the Mini-SAR sends pulses of left polarised radio waves out to measure the surface roughness of the moon. While smooth surfaces send back a reversed right polarised wave, rough areas return left polarised waves. Ice, which is transparent to radio waves, also sends back left polarised waves. The Mini-SAR measures the ratio of left to right circular polarised power sent back, or the circular polarised ratio (CPR). However, a high CPR alone can not distinguish between rough patches and regions with ice. The North Pole craters had a high CPR on the inside with a low CPR on the edges. This result suggests that a material enclosed in the craters rather than surface roughness caused the high CPR signal. According to NASA, the ice should be relatively pure and at least a meter thick to give such signature.

Regolith samples of the Moon covering, which were transported to Earth by several Apollo missions, had been investigated thoroughly in laboratories using conventional methods [98-99]. Gas adsorption isotherms revealed small specific surface areas below 1 $m^2\ g^{-1}$. A water vapour isotherm is depicted in Figure 29. The material is rather hydrophobic with small surface area and can hardly store water.

By means of mass spectrometry, traces of water were found in the interior of volcanic pearls from the Moon, which had been brought to Earth by Apollo

15 [96, 100]. Assuming that 95 % of the water introduced vaporised immediately, the content within Moon's magma is estimated to about 750 ppm. On the other hand, the comparatively large occurrence of the element chlorine suggest, that it is really dry on Moon.

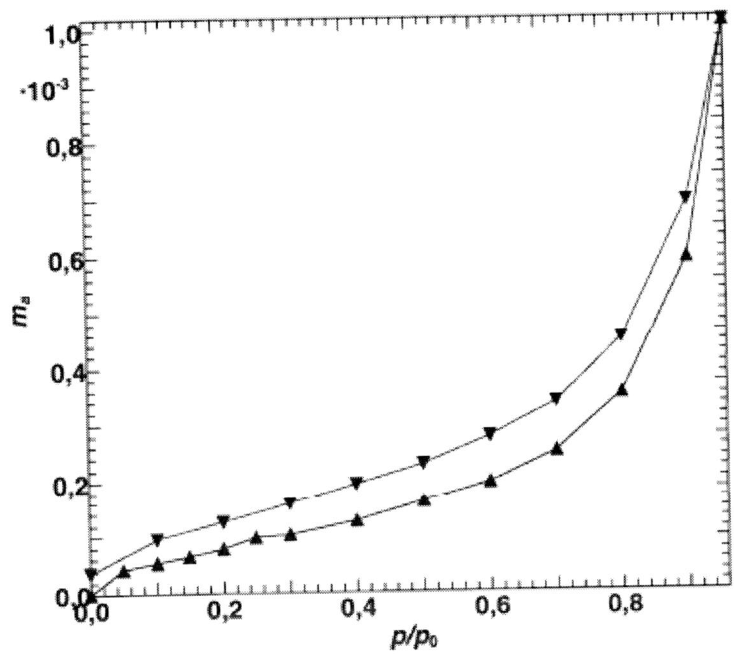

Figure 28. Water vapour adsorption and desorption isotherm at 24,9 C at Lunar regolith sample 12001.922 from the Apollo 12 mission. ▲ adsorption, ▼ desorption.

Mars

Most of the water is locked as permafrost in soil and as polar caps together with carbon dioxide ice [101]. Though the temperatures at the Mars fluctuate between -55 and 27 °C, there are no bodies of liquid water, which could create a hydrosphere. Occasionally, droplets at stones have been observed, most probably solutions of perchlorate. Only a small amount of water vapour is in the atmosphere. Water from the poles generates cirrus clouds by sublimation and fog is observed in some valleys in the morning. During summer the northern region of the North Pole is covered with a mixture of water ice and carbon dioxide ice that is 1000 km in diameter and about 5 km thick. In the south-polar region, layered deposits, which are 350 km in diameter and 1.5 km thick, with large amounts of water ice could be identified. Below these caps, deposits in soil, which could cover -when melted- the planet with an 11 m thick water layer, have been detected. For comparison, on Earth a similar coverage would be 3 km thick. In addition, deposits in the soil in the form of ice stocks or permafrost are confirmed for each region of Mars

The water occurrence on Mars was investigated during several satellite missions, first by the Odyssey mission in 2002 [102-103]. Galactic cosmic ray particles react with matter of the planets surface, and they are slowed by loss of energy. The spectra of the re-emitted neutron and gamma rays are evaluated with regard to hydrogen and oxygen peaks, which can be attributed to water at the surface down to some depth. The south polar deposits have been detected by lowered radar echo.

7. Standardization

Scientific problems concerning weather and climate are handled by the World Meteorological Organization (WMO), which is an intergovernmental organization and became the specialised agency of the United Nations for meteorology, operational hydrology and related geophysical sciences. The WMO has its headquarters in Geneva, Switzerland. There exist also national administrations and scientific societies in most countries.

The commonly used term "humidity" includes also liquid contents other than water, solvents e.g. [58]. Indeed, gravimetric methods merely measure the sum of volatile substances within a solid whereas, in general, the task is often to determine the water content only. Therefore, standardisation of measuring methods and procedures is required [2, 7, 19]. Several working groups of national and international standardisation committees develop standards of humidity measurements and related items. One reason is that new measuring instruments, which are in competition to standardised but obsolete methods, are offered. Because water adsorption largely depends on the solid partner, measuring procedures are often specialised. More than 200 standards exist for the investigation of about 70 different materials, like building materials, soil, paint, ceramics, coal, ores, plastics, food, leather, stones, paper and paperboard. More than 50 standards are already harmonised in international ISO standards. In addition, manufacturers as well as vocational and scientific societies have developed special testing specifications [26-27, 104-108]. A working committee of the German standardisation organisation DIN has just released a basic standard on gravimetric humidity measurements [109]. Certified methods of the measurement of air humidity are compiled in a vade-mecum in Germany [110]. The principle of thermal analysis is described in DIN 51006/7 [111-112].

Calibration of instruments is made by Services' in the USAPhysikalisch-Technische Bundesanstalt' and 'Deutscher Kalibrierdienst' , by 'National Physical Laboratory' and 'United Kingdom Accreditation Service' in the UK e.g. A Gravimetric Hygrometer is considered to be a primary source for calibration. National standards have been developed in US, UK, EU and Japan.

Two thermometers can be compared by immersing them both in an insulated vessel of water and stirring vigorously to minimize temperature variations. If handled with care a high-quality liquid-in-glass thermometer should remain stable for some years. Hygrometers must be calibrated in air, which is a much less effective heat transfer medium than water, and many types are subject to drift and need regular recalibration. A further difficulty is that most hygrometers sense relative humidity rather than the absolute amount of water present. But the relative humidity is a function of temperature and absolute moisture content, so that small temperature variations within the air of a test chamber are translated into relative humidity variations.

Standards for the measurement of the water content are listed in Table 4. A list of national identifiers for standards and the assigned standardisation bodies are shown in Table 5. Symbols used are reported in Table 6.

Table 4. Standards for the measurement of water content, dry mass or water sorption

ASTM		ASTM International (former American Society for Testing and Materials), 100 Barr Harbor Drive, West Conshohocken, PA, USA
C 70	06	Test method for surface moisture in fine aggregate.
C 121	09	Test method for water absorption of slate.
C 156	09	Test method for water loss [from a mortar specimen] through liquid membrane-forming curing compounds for concrete.
C 272	91	Test method for water absorption of core materials for structural sandwich constructions.
C 324	01	Test method for free moisture in ceramic whiteware clays.
C 370	88	Test method for moisture expansion of fired whiteware products.
C 373	88	Test method for water absorption, bulk density, apparent porosity, and apparent specific gravity of fired whiteware products.
C 562	91	Test method for moisture in a graphite sample.

Table 4. (Continued)

ASTM		ASTM International (former American Society for Testing and Materials), 100 Barr Harbor Drive, West Conshohocken, PA, USA
C 566	97	Test method for total evaporable moisture content of aggregate by drying.
C 1104	00	Test method for determining the water vapor sorption of unfaced mineral fiber insulation.
C 1258	08	Test method for elevated temperature and humidity resistance of vapor retarders for insulation.
C 1616	07	Test method for determining the moisture content of organic and inorganic insulation materials by weight.
C 1699	09	Test method for moisture retention curves of porous building materials using pressure plates.
D 280	01	Test methods for hygroscopic moisture (and other matter volatile under the test conditions) in pigments.
D 425	88	Test method for centrifuge moisture equivalent of soils.
D 558	04	Test methods for moisture-density (unit weight) relations of soil-cement mixtures.
D 570	95	Test method for water absorption of plastics.
D 644	99	Test method for moisture content of paper and paperboard by oven drying.
D 890	98	Test method for water in liquid naval stores.
D 1123	99	Test methods for water in engine coolant concentrate by the Karl Fischer reagent method.
D 1151	00	Practice for effect of moisture and temperature on adhesive bonds.
D 1193	06	Specification for reagent water.
D 1348	94	Test methods for moisture in cellulose.
D 1364	02	Test method for water in volatile solvents (Karl Fischer reagent titration method).
D 1412	07	Test method for equilibrium moisture of coal at 96 to 97 % relative humidity and 30 °C.
D 1461	85	Test method for moisture or volatile distillates in bituminous paving mixtures.
D 1533	00	Test method for water in insulating liquids by coulometric Karl Fischer titration.
D 1558	99	Test method for moisture content penetration resistance relationships fine-grained soils.
D 1576	90	Test method for moisture in wool by oven-drying.
D 1631	10	Test method for water in phenol and related materials by the iodine reagent method.
D 1640	06	Test methods for drying, curing, or film formation of organic coatings at room temperature.
D 1815	00	Test method for water absorption (static) of vegetable tanned leather.
D 1860	95	Test method for moisture and creosote-type preservative in wood.

ASTM		ASTM International (former American Society for Testing and Materials), 100 Barr Harbor Drive, West Conshohocken, PA, USA
D 1864	89	Test method for moisture in mineral aggregate used on built-up roofs.
D 1909	04	Table of commercial moisture regains for textile fibres.
D 2118	05	Practice for assigning a standard commercial moisture content for wool and it's products.
D 2132	03	Test method for dust-and-fog tracking and erosion resistance of electrical insulating materials.
D 2216	05	Test method for laboratory determination of water (moisture) content of soil and rock by mass.
D 2247	02	Practice for testing water resistance of coatings in 100 % relative humidity.
D 2462	90	Test method for moisture in wool by distillation with toulene.
D 2495	07	Test method for moisture in cotton by oven-drying.
D 2525	90	Practice for sampling wool for moisture.
D 2713	07	Test method for dryness of propane (valve-freeze method).
D 2842	06	Test method for water absorption of rigid cellular plastics.
D 2867	09	Test methods for moisture in activated carbon.
D 2944	71	Test methods of sampling processed peat materials.
D 2961	02	Test method for single-stage total moisture less than 15 % in coal reduced to 2.36 mm (No. 8 sieve) topsize.
D 2974	07	Test methods for moisture, ash, and organic matter of peat and other organic soils.
D 2980	04	Test method for volume mass, moisture-holding capacity, and porosity of saturated peat materials.
D 2987	88	Test method for moisture content of asbestos fiber.
D 3173	03	Test method for moisture in the analysis sample of coal and coke.
D 3175	07	Test method for volatile matter in the analysis sample of coal and coke.
D 3201	08	Test method for hygroscopic properties of fire-retardant wood and wood-based products.
D 3285	93	Test method for water absorptiveness of nonbibulous paper and paperboard (Cobb test).
D 3302	10	Test method for total moisture in coal.
D 3401	97	Test methods for water in halogenated organic solvents and their admixtures.
D 3790	79	Test method for volatile matter (moisture) of leather by oven drying.
D 3816	93	Test method for water penetration rate of pressure-sensitive tapes.
D 4017	02	Test method for water in paints and paint materials by Karl Fischer method.
D 4230	02	Test method of measuring humidity with cooled-surface condensation (dew-point) hygrometer.
D 4263	83	Standard test method for indicating moisture in concrete by the plastic sheet method

Table 4. (Continued)

ASTM		ASTM International (former American Society for Testing and Materials), 100 Barr Harbor Drive, West Conshohocken, PA, USA
D 4377	00	Test method for water in crude oils by potentiometric Karl Fischer titration.
D 4442	07	Test methods for direct moisture content measurement of wood and wood-base materials.
D 4444	08	Test method for laboratory standardization and calibration of hand-held moisture meters.
D 4502	92	Test method for heat and moisture resistance of wood-adhesive joints.
D 4643	08	Test method for determination of water (moisture) content of soil by microwave oven heating.
D 4672	00	Test methods for polyurethane raw materials: Determination of water content of polyols.
D 4772	09	Test method for surface water absorption of terry fabrics (water flow).
D 4902	99	Test method for evaporation and drying of analytical solutions.
D 4928	00	Test methods for water in crude oils by coulometric Karl Fischer titration.
D 4931	06	Test method for gross moisture in green petroleum coke.
D 4933	99	Guide for moisture conditioning of wood and wood-base materials.
D 4942	89	Test methods for water pickup of lithographic printing inks and vehicles in a laboratory mixer.
D 4944	04	Test method for field determination of water (moisture) content of soil by the calcium carbide gas pressure tester.
D 4959	07	Test method for determination of water (moisture) content of soil by direct heating.
D 5032	97	Practice for maintaining constant relative humidity by means of aqueous glycerin solutions.
D 5229	92	Test method for moisture absorption properties and equilibrium conditioning of polymer matrix composite materials.
D 5348	95	Test method for determination of the moisture content of sulfonated and sulfated oils by distillation with xylene.
D 5349	95	Test method for determination of the moisture and volatile content of sulfonated and sulfated oils by hot-plate method.
D 5454	04	Test method for water vapor content of gaseous fuels using electronic moisture analyzers.
D 5455	93	Test method for short-term liquid sorption into paper (Bristow test).
D 5556	95	Test method for determination of the moisture and other volatile matter contained in fats and oils used in fat liquors and softening compounds.
D 5460	02	Test method for rubber compounding materials - Water in rubber additives.
D 5530	94	Test method for total moisture of hazardous waste fuel by Karl Fischer titrimetry.
D 5637	05	Test method for moisture resistance of electrical insulating varnishes.

ASTM		ASTM International (former American Society for Testing and Materials), 100 Barr Harbor Drive, West Conshohocken, PA, USA
D 5744	07	Test method for laboratory weathering of solid materials using a humidity cell.
D 5895	08	Test methods for evaluating drying or curing during film formation of organic coatings using mechanical recorders.
D 6031	96	Test method for logging in situ moisture content and density of soil and rock by the nuclear method in horizontal, slanted, and vertical access tubes.
D 6304	07	Test method for determination of water in petroleum products, lubricating oils, and additives by coulometric Karl Fisher titration.
D 6403	99	Test method for determining moisture in raw and spent materials.
D 6489	99	Test method for determining the water absorption of hardened concrete treated with a water repellent coating.
D 6565	00	Test method for determination of water (moisture) content of soil by the time-domain reflectometry (TDR) method.
D 6642	01	Guide for comparison of techniques to quantify the soil-water (moisture) flux.
D 6658	01	Test method for volatile matter (moisture) of wet blue by oven drying.
D 6782	05	Test methods for standardization and calibration of in-line dry lumber moisture meters.
D 6869	03	Test method for coulometric and volumetric determination of moisture in plastics using the Karl Fischer reaction (the reaction of iodine with water).
D 6980	09	Test method for determination of moisture in plastics by loss in weight.
D 7013	04	Guide for nuclear surface moisture and density gauge calibration facility setup.
D 7191	10	Test method for determination of moisture in plastics by relative humidity sensor.
D 7404	07	Test method for determination of emulsified asphalt residue by moisture analyzer.
D 7438	08	Practice for field calibration and application of hand-held moisture meters.
D 7546	09	Test method for determination of moisture in new and in-service lubricating oils and additives by relative humidity sensor.
E 104	02	Practice for maintaining constant relative humidity by means of aqueous solutions.
E 203	08	Test method for water using volumetric Karl Fischer titration.
E 337	02	Test method for measuring humidity with a psychrometer (the measurement of wet- and dry-bulb temperatures).
E 398	03	Test method for water vapor transmission rate of sheet materials using dynamic relative humidity measurement.
E 410	08	Test method for moisture and residue in liquid chlorine.

Table 4. (Continued)

ASTM		ASTM International (former American Society for Testing and Materials), 100 Barr Harbor Drive, West Conshohocken, PA, USA
E 790	08	Test method for residual moisture in a refuse-derived fuel analysis sample.
E 871	82	Test method for moisture analysis of particulate wood fuels.
E 949	88	Test method for total moisture in a refuse-derived fuel laboratory sample.
E 1064	08	Test method for water in organic liquids by coulometric Karl Fischer titration.
E 1358	97	Test method for determination of moisture content of particulate wood fuels using a microwave oven.
E 1868	09	Test method for loss-on-drying by thermogravimetry.
E 255	07	Test method for humidity calibration (or conformation) of humidity generators for use with thermogravimetric analyzers.
F 1397	93	Test method for determination of moisture contribution by gas distribution system components.
F 1869	10	Test method for measuring moisture vapor emission rate of concrete subfloor using anhydrous calcium chloride.
F 2170	09	Test method for determining relative humidity in concrete floor slabs using in situ probes.
F 2420	05	Test method for determining relative humidity on the surface of concrete floor slabs using relative humidity probe measurement and insulated hood.
F 2298	03	Test methods for water vapor diffusion resistance and air flow resistance of clothing materials using the dynamic moisture permeation cell.
G 60	01	Practice for conducting cyclic humidity tests.

BS		BSI British Standards Institution, 389 Chiswick High Road, London, W4 4AL, U.K.
BS 1016-104.1	99	Method for analysis and testing of coal and coke - Proximate analysis - Determination of moisture content of the general analysis test sample.
1377-2	90	Methods of test for soils for civil engineering purposes - Classification tests.
1377-4	90	Methods of test for soils for civil engineering purposes - Part 4: Compaction-related tests.
1881-128	97	Testing concrete – Part 128. Methods for analysis of fresh concrete
1924-1	90	Stabilized materials for civil engineering purposes - General requirements, sampling, sample preparation and tests on materials before stabilization.
1924-2	90	Stabilized materials for civil engineering purposes - Methods of test for cement-stabilized and limestabilized materials.

ASTM		ASTM International (former American Society for Testing and Materials), 100 Barr Harbor Drive, West Conshohocken, PA, USA
2782-11.1103C	98	≡ EN 12118: Plastics piping systems - Determination of moisture content in thermoplastics by coulometry.
2880	91	Methods of testing cellulose acetate flake.
2975	04	Sampling and analysis of glass-making sands - Part 1: Methods for sampling and physical testing of glassmaking sands.
3482-2	91	Methods of test for desiccants - Determination of moisture content.
3823	90	Grading of ash and hickory wood handles for hand tools.
3958-1	82	Thermal insulating materials - Magnesia preformed insulation.
3958-2	82	Thermal insulating materials - Calcium silicate preformed insulation.
3958-3	85	Thermal insulating materials - Metal mesh faced manmade mineral fibre mattresses.
3958-4	82	Thermal insulating materials - Bonded preformed manmade mineral fibre pipe sections.
3986-1	98	Methods of test for drying performance of agricultural grain dryers - General procedures and limited requirements for wheat.
3986-2	98	Methods of test for drying performance of agricultural grain dryers - Additional procedures and crop specific requirements.
4317-15	91	Methods of test for cereals and pulses - Determination of moisture content of maize (milled and whole).
4317-2	87	Methods of test for cereals and pulses - Determination of moisture content of cereals and cereal products (Reference method).
4317-22	89	Methods of test for cereals and pulses - Determination of water absorption of flour and rheological properties of doughs using a valorigraph.
4317-25	90	Methods of test for cereals and pulses – determination of wet gluten content in wheat flour by mechanical means.
4401-3	97	Methods of test for meat and meat products - determination of moisture content (Reference method).
4585-2	82	Methods of test for spices and condiments - Determination of moisture content (Entrainment method).
4993-5	80	Methods of test for aluminium fluoride for industrial use - Determination of moisture content (Karl Fischer method).
4993-6	80	Methods of test for aluminium fluoride for industrial use - Determination of moisture content (Gravimetric method).
5050-9(	80	Methods of test for cryolite - Determination of moisture content (Karl Fischer method).
5050-10	80	Methods of test for cryolite - Determination of moisture content (Gravimetric method).
5072-2	74	Methods of test for sodium fluoride for industrial use - deTermination of moisture content.

Table 4. (Continued)

ASTM		ASTM International (former American Society for Testing and Materials), 100 Barr Harbor Drive, West Conshohocken, PA, USA
5268-2	02	Structural use of timber - Part 2: Code of practice for permissible stress design, materials and workmanship.
5309-2	76	Methods for sampling chemical products - Sampling of gases.
5350-C5	02	Methods of test for adhesives - Determination of bond strength in longitudinal shear for rigid adherends.
5752-1	79	Methods of test for coffee and coffee products – Green coffee - Determination of moisture content (Basic reference method).
5752-2	84	Methods of test for coffee and coffee products – Green coffee: determination of moisture content (Routine method).
5752-13	95	Methods of test for coffee and coffee products – Roasted ground coffee - Determination of moisture content - Karl Fischer method (Reference method).
5752-14	95	Methods of test for coffee and coffee products – Roasted ground coffee - Determination of moisture content (loss in mass at 103 °C, routine method).
6091-2	87	Vulcanized fibre for electrical purposes - Methods of test.
6986-3	90	Analysis of instant tea - Method for determination of moisture content (loss in mass at 103 °C).
7312	90	Methods of test for magnetite for use in coal preparation.
7319-2	90	Analysis of sodium chloride for industrial use – Method for determination of moisture content.
7970	05	Electric cables - Metallic wire and foil sheath constructions of power cables having XPLE insulation for voltage from 66 kV (U[M] = 72.5 kV) to 132 kV (U[M] = 145 kV).

BVL		Bundesamt für Verbraucherschutz und Lebensmittelsicherheit, Bundesallee 50, 38116 Braunschweig (Germany)
57.20.19-1(EG)	82	Determination of the drying loss of sodium nitrite (E250).

DIN		Deutsches Institut für Normung e.V., Burggrafenstraße 6, D-10772 Berlin, Germany
4051	02	Sewer clinkers - requirements, testing, quality control.
8063-5	99	Pipe joints and components of unplasticized poly(vinyl chloride) (PVC-U) for pipes under pressure - Part 5: General quality requirements, testing.
8949	00	Refrigerant filter driers – Testing.
1048-1	91	Testing methods for concrete; fresh concrete
1948-5	91	Testing methods for concrete; hardened concret, specially prepared specimens

ASTM		ASTM International (former American Society for Testing and Materials), 100 Barr Harbor Drive, West Conshohocken, PA, USA
10236	01	Analysis of spices and condiments - Determination of loss in mass of capsicum and allium species and of dried vegetables by vacuum oven drying.
10304-2	69	Determination of moisture content and dry substance of glucose syrup; Vacuum oven method.
10304-3	71	Determination of moisture content and dry matter content of glucose syrup; Refractive index method.
10321	80	Determination of the water content of dried milk.
10360	72	Fruit and vegetables - Determination of dry matter content of deep-frozen or heat-preserved spinach.
10453	83	Determination of water content of caseins and caseinates - reference method.
18132	95	Soil, testing procedures and testing equipment - Determination of water absorption.
18180	89	Gypsum plasterboards - Types and requirements.
19682-5	07	Soil quality - Field tests - Part 5: Determination of soil moisture.
19747	09	Investigation of solids - Pre-treatment, preparation and processing of samples for chemical, biological and physical investigations.
29597	95	Aerospace - Application of liquid coating materials.
43903	87	Electrical switch gear and control gear; Humidity of compressed air; Physical relationships, methods of measuring.
45922-1	84	≡ CECC 41000 Harmonized system of quality assessment for electronic components; Generic specification: Potentiometers.
46449	70	Insulating varnishes and insulating resin materials for electrical purposes; Finishing varnishes, Test methods.
51078	02	Testing of ceramic raw materials and materials - Preparation of samples for the determination of change of mass during drying and for chemical analysis.
52251-2	87	Indirect methods of determining the frost resistance of roofing tiles; Determination of water absorption.
52251-3	87	Indirect methods of determining the frost resistance of roofing tiles; Determination of coefficient of impregnation.
52331	95	Testing of glass; Crushing and drying of samples for chemical analysis.
52459	81	Testing of building sealants; Determination of the water absorption of backfill material; Retentive capacity.
53129	01	Testing of solid fibre board - Water absorption and swelling - Water immersion.
53814	74	Testing of textiles; Determination of the water retention power of fibres and yarn cuttings.
53923	78	Testing of textiles; Determination of water absorption of textile fabrics.

Table 4. (Continued)

ASTM		ASTM International (former American Society for Testing and Materials), 100 Barr Harbor Drive, West Conshohocken, PA, USA
54540-9	07	Testing of hygiene products - Tissue and nonwoven - Part 9: Determination of moisture content.
57291-2	79	≡ VDE 0291-2 Compounds for use in cable accessories; Casting resinous compounds before cure and in the cured state.
65046-2	98	Aerospace - Methods of test for surface protective coatings; paints, varnishes - Part 2: Specimens, preparation of specimens, drying.
66138	08	Isothermal measurement of the sorption of vapours at solids.
DIN V		Deutsches Institut für Normung e.V., Burggrafenstraße 6, D-10772 Berlin, Germany
52049	98	Testing of bitumen - Determination of water content of bituminous emulsions - Drying test.
EN		Comité Européenne de Normalisation - CEN, Rue de Stassart 36, B-1050 Bruxelles, Belgium Note: Once adopted, EN standards can only be purchased in the form of national adoptions. A list of national identifiers for standards and the assigned standardization bodies is given at the end of this table.
3-7	04	Portable fire extinguishers - Part 7: Characteristics, performance requirements and test methods. [national adoptions from EN: BS DIN I.S. NBN NEN NF NS ÖNORM PN SN SS UNE UNI]
62	77	Glass reinforced plastics - Standard atmospheres for conditioning and testing. [national adoption from EN: UNI]
121	91	Extruded ceramic tiles with low water absorption (E <= 3%) - Group A1. [national adoptions from EN: I.S. SN]
159	91	Dust-pressed ceramic tiles with water absorption E > 10% - Group B3. [national adoptions from EN: I.S. SN]
176	92	Dust-pressed ceramic tiles with a low water absorption (E <= 3%) - Group B1. [national adoptions from EN: I.S. ÖNORM SN]
177	91	Dust-pressed ceramic tiles with a water absorption of 3% < E <= 6% - Group B2A. [national adoptions from EN: I.S. SN]
178	91	Dust-pressed ceramic tiles with a water absorption of 6% < E <= 10% - Group B2B. [national adoptions from EN: I.S. ÖNORM SN]

ASTM		ASTM International (former American Society for Testing and Materials), 100 Barr Harbor Drive, West Conshohocken, PA, USA
186-1	91	Extruded ceramic tiles with a water absorption of 3% < E <= 6% (Group A2A); Part 1. [national adoptions from EN: I.S. SN]
186-2	91	Extruded ceramic tiles with a water absorption of 3% < E <= 6% (Group A2A) - Part 2. [national adoptions from EN: I.S. PN SN]
187-1	91	Ceramic tiles - Extruded ceramic tiles with a water absorption of 6% < E <= 10% (Group A2B) - Part 1. [national adoptions from EN: I.S. SN]
187-2	91	Ceramic tiles - Extruded ceramic tiles with a water absorption of 6% < E <= 10% (Group A2B) - Part 2. [national adoptions from EN: I.S. ÖNORM SN]
188	91	Ceramic tiles - Extruded ceramic tiles with a water absorption of E > 10% (Group A3). [national adoptions from EN: I.S. SN]
212	03	≡ UNE 56424: Wood preservatives - General guidance on sampling and preparation for analysis of wood preservatives and treated timber. [national adoptions from EN: BS DIN I.S. NBN NEN NF NS ÖNORM PN SN SS UNE UNI]
322	93	Wood-based panels - Determination of moisture content. [national adoptions from EN: BS DIN I.S. NBN NEN NF NS ÖNORM PN SN SS UNE UNI]
382-1	94	Fibreboards - Determination of surface absorption - Part 1: Test method for dry process fibreboards. [national adoptions from EN: BS DIN I.S. NBN NEN NF NS ÖNORM PN SN SS UNE UNI]
382-2	94	Fibreboards - Determination of surface absorption - Part 2: Test methods for hardboards. [national adoptions from EN: BS DIN I.S. NBN NEN NF NS ÖNORM PN SN SS UNE UNI]
480-5	05	Admixtures for concrete, mortar and grout - Test methods - Part 5: Determination of capillary absorption. [national adoptions from EN: BS DIN I.S. NBN NEN NF NS ÖNORM PN SN SS UNE UNI]
771-1	03	Specification for masonry units – Part 1: Clay masonry units. [national adoptions from EN: BS DIN I.S. NBN NEN NF NS ÖNORM PN SN SS UNE UNI]
771-2	03	Specification for masonry units – Part 2: Calcium silicate masonry units. [national adoptions from EN: BS DIN I.S. NBN NEN NF NS ÖNORM PN SN SS UNE UNI]

Table 4. (Continued)

ASTM		ASTM International (former American Society for Testing and Materials), 100 Barr Harbor Drive, West Conshohocken, PA, USA
771-3	03	Specification for masonry units – Part 3: Aggregate concrete masonry units (dense and light-weight aggregates). [national adoptions from EN: BS DIN I.S. NBN NEN NF NS ÖNORM PN SN SS UNE UNI]
771-4	03	Specification for masonry units – Part 4: Autoclaved aerated concrete masonry units. [national adoptions from EN: BS DIN I.S. NBN NEN NF NS ÖNORM PN SN SS UNE UNI]
771-5	03	Specification for masonry units – Part 5: Manufactured stone masonry units. [national adoptions from EN: BS DIN I.S. NBN NEN NF NS ÖNORM PN SN SS UNE UNI]
771-6	05	Specification for masonry units - Part 6: Natural stone masonry units. [national adoptions from EN: BS DIN I.S. NBN NEN NF NS ÖNORM PN SN SS UNE UNI]
772-7	98	Methods of test for masonry units - Part 7: Determination of water absorption of clay masonry damp proof course units by boiling in water. [national adoptions from EN: BS DIN I.S. NBN NEN NF NS ÖNORM PN SN UNE UNI]
772-10	99	Methods of test for masonry units - Part 10: Determination of moisture content of calcium silicate and autoclaved aerated concrete units. [national adoptions from EN: BS DIN I.S. NBN NEN NF NS ÖNORM PN SN UNE UNI]
772-11	00	Methods of test for masonry units - Part 11: Determination of water absorption of aggregate concrete, manufactured stone and natural stone masonry units due to capillary action and the initial rate of water absorption of clay masonry units. [national adoptions from EN: BS DIN I.S. NBN NEN NF NS ÖNORM PN SN SS UNE UNI]
844-4	97	Round and sawn timber - Terminology - Part 4: Terms relating to moisture content. [national adoptions from EN: BS DIN I.S. NBN NEN NF NS ÖNORM PN SN SS UNE UNI]
993-1	95	Methods of test for dense shaped refractory products - Part 1: Determination of bulk density, apparent porosity and true porosity. [national adoptions from EN: BS DIN I.S. NBN NEN NF NS ÖNORM PN SN SS UNE UNI]

ASTM		ASTM International (former American Society for Testing and Materials), 100 Barr Harbor Drive, West Conshohocken, PA, USA
1015-10	99	Methods of test for mortar for masonry - Part 10: Determination of dry bulk density of hardened mortar. [national adoptions from EN: BS DIN I.S. NBN NEN NF NS ÖNORM PN SN SS UNE UNI]
1015-18	02	Methods of test for mortar for masonry - Part 18: Determination of water absorption coefficient due to capillary action of hardened mortar. [national adoptions from EN: BS DIN I.S. NBN NEN NF NS ÖNORM PN SN SS UNE UNI]
1015-19	98	Methods of test for mortar for masonry - Part 19: Determination of water vapour permeability of hardened rendering and plastering mortars. [national adoptions from EN: BS DIN I.S. NBN NEN NF NS ÖNORM PN SN SS UNE UNI]
1097-6	00	Tests for mechanical and physical properties of aggregates - Part 6: Determination of particle density and water absorption. [national adoptions from EN: BS DIN I.S. NBN NEN NF NS ÖNORM PN SN SS UNE UNI]
1097-10	99	Tests for mechanical and physical properties of aggregates - Determination of water suction height. [national adoptions from EN: BS DIN I.S. NBN NEN NF NS ÖNORM PN SN SS UNE UNI]
1161	96	Feather and down - Test methods - Determination of moisture content. [national adoptions from EN: BS DIN I.S. NBN NEN NF NS ÖNORM PN SN SS UNE UNI]
1170-6	97	Precast concrete products - Test method for glass-fibre reinforced cement - Part 6: Determination of the absorption of water by immersion and determination of the dry density. [national adoptions from EN: BS DIN I.S. NBN NEN NF NS ÖNORM PN SN SS UNE UNI]
1217	98	Materials and articles in contact with foodstuffs - Test methods for absorption of ceramic articles. [national adoptions from EN: BS DIN I.S. NBN NEN NF NS ÖNORM PN SN SS UNI]
1348	07	Adhesives for tiles - Determination of tensile adhesion strength for cementitious adhesives. [national adoptions from EN: BS DIN I.S. NBN NEN NF NS ÖNORM PN SN SS UNE UNI]
1353	96	Determination of moisture content of autoclaved aerated concrete. [national adoptions from EN: BS DIN I.S. NBN NEN NF NS ÖNORM PN SN SS UNE UNI]

Table 4. (Continued)

ASTM		ASTM International (former American Society for Testing and Materials), 100 Barr Harbor Drive, West Conshohocken, PA, USA
1402-3	03	Unshaped refractory products - Part 3: Characterization as received. [national adoptions from EN: BS DIN I.S. NBN NEN NF NS ÖNORM PN SN SS UNE UNI]
1609	96	Thermal insulating products for building applications - Determination of short term water absorption by partial immersion. [national adoptions from EN: BS DIN GOST R I.S. NBN NEN NF NS ÖNORM PN SN SS UNE UNI]
1860-2	05	Appliances, solid fuels and firelighters for barbecueing - Part 2: Barbecue charcoal and barbecue charcoal briquettes - Requirements and test methods. [national adoptions from EN: BS DIN I.S. NBN NEN NF NS ÖNORM PN SN SS UNE UNI]
1925	99	Natural stone test methods - Determination of water absorption coefficient by capillarity. [national adoptions from EN: BS DIN I.S. NBN NEN NF NS ÖNORM PN SN SS UNE UNI]
2155-2	93	Test methods for transparent materials for aircraft glazing - Determination of water absorption. [national adoptions from EN: BS DIN I.S. NBN NEN NF NS ÖNORM PN SNV DIN SS UNE UNI]
2155-13	93	Test methods for transparent materials for aircraft glazing - Part 13: Determination of temperature at deflection under load. [national adoptions from EN: BS DIN I.S. NBN NEN NF NS ÖNORM PN SNV DIN SS UNE UNI]
2378	95	Aerospace series - Fibre reinforced plastics - Determination of water absorption by immersion. [national adoptions from EN: BS DIN I.S. NBN NEN NF NS ÖNORM PN SNV DIN SS UNE UNI]
2667-6	01	Aerospace series - Non-metallic materials - Foaming structural adhesives - Test methods - Part 6: Determination of water absorption. [national adoptions from EN: BS DIN I.S. NBN NEN NF NS ÖNORM PN SN SS UNI]
12048	96	Solid fertilizers and liming materials - Determination of moisture content - Gravimetric method by drying at (105 ± 2)°C. [national adoptions from EN: BS DIN I.S. NBN NEN NF NS ÖNORM PN SN SS UNE UNI]
12049	96	Solid fertilizers and liming materials - Determination of moisture content - Gravimetric method by drying under reduced pressure. [national adoptions from EN: BS DIN I.S. NBN NEN NF ÖNORM SN SS UNE UNI]

ASTM		ASTM International (former American Society for Testing and Materials), 100 Barr Harbor Drive, West Conshohocken, PA, USA
12087	97	Thermal insulating products for building applications - Determination of long term water absorption by immersion. [national adoptions from EN: BS DIN GOST R I.S. NBN NEN NF NS ÖNORM PN SN SS UNE UNI]
12088	97	Thermal insulating products for building applications - Determination of long term water absorption by diffusion. [national adoptions from EN: BS DIN I.S. NBN NEN NF NS ÖNORM PN SN SS UNE UNI]
12105	98	Resilient floor coverings - Determination of moisture content of agglomerated composition cork. [national adoptions from EN: BS DIN I.S. NBN NEN NF NS ÖNORM PN SN SS UNE UNI]
12118	97	Plastics piping systems - Determination of moisture content in thermoplastics by coulometry. [national adoptions from EN: BS DIN I.S. NBN NEN NF NS ÖNORM PN SN SS UNE UNI]
12390-8	09	Testing hardened concrete - Part 8: Depth of penetration of water under pressure. [national adoptions from EN: BS DIN I.S. NBN NEN NF NS ÖNORM PN SN SS UNE UNI]
12429	98	Thermal insulating products for building applications - Conditioning to moisture equilibrium under specified temperature and humidity conditions. [national adoptions from EN: BS DIN I.S. NBN NEN NF NS ÖNORM PN SN SS UNE UNI]
12509	01	Timber poles for overhead lines - Test methods - Determination of modulus of elasticity, bending strength, density and moisture content. [national adoptions from EN: BS DIN I.S. NBN NEN NF NS ÖNORM PN SN SS UNE UNI]
12746	00	Footwear - Test methods for insoles and insocks - Water absorption and desorption. [national adoptions from EN: BS DIN I.S. NBN NEN NF NS ÖNORM PN SN SS UNE UNI]
13040	07	Soil improvers and growing media - Sample preparation for chemical and physical tests, determination of dry matter content, moisture content and laboratory compacted bulk density. [national adoptions from EN: BS DIN I.S. NBN NEN NF NS ÖNORM PN SN SS UNE UNI]

Table 4. (Continued)

ASTM		ASTM International (former American Society for Testing and Materials), 100 Barr Harbor Drive, West Conshohocken, PA, USA
13183-1	02	Moisture content of a piece of sawn timber - Determination by oven dry method. [national adoptions from EN: BS DIN I.S. NBN NEN NF NS ÖNORM PN SN SS UNE UNI]
13183-2	02	Moisture content of a piece of sawn timber - Estimation by electrical resistance method. [national adoptions from EN: BS DIN I.S. NBN NEN NF NS ÖNORM PN SN SS UNE UNI]
13183-3	05	Moisture content of a piece of sawn timber - Part 3: Estimation by capacitance method. [national adoptions from EN: BS DIN I.S. NBN NEN NF NS ÖNORM PN SN SS UNE UNI]
13407	06	Wall-hung urinals - Functional requirements and test methods. [national adoptions from EN: BS DIN I.S. NBN NEN NF NS ÖNORM PN SN SS UNE UNI]
13579	02	Products and systems for the protection and repair of concrete structures - Test methods - Drying test for hydrophobic impregnation. [national adoptions from EN: BS DIN I.S. NBN NEN NF NS ÖNORM PN SN SS UNE UNI]
13748-1	04	Terrazzo tiles - Part 1: Terrazzo tiles for internal use. [national adoptions from EN: BS DIN I.S. NBN NEN NF NS ÖNORM PN SN SS UNE UNI]
13748-2	04	Terrazzo tiles - Part 2: Terrazzo tiles for external use. [national adoptions from EN: BS DIN I.S. NBN NEN NF NS ÖNORM PN SN SS UNE UNI]
13755	02	Natural stone test methods - Determination of water absorption at atmospheric pressure. [national adoptions from EN: BS DIN I.S. NBN NEN NF NS ÖNORM PN SN SS UNE UNI]
13815	06	Fibrous gypsum plaster casts - Definitions, requirements and test methods. [national adoptions from EN: BS DIN I.S. NBN NEN NF NS ÖNORM PN SN SS UNI]
14251	03	Structural round timber - Test methods. [national adoptions from EN: BS DIN I.S. NBN NEN NF NS ÖNORM PN SN SS UNE UNI]
14298	04	Sawn timber - Assessment of drying quality. [national adoptions from EN: BS DIN I.S. NBN NEN NF NS ÖNORM PN SN SS UNE UNI]

ASTM		ASTM International (former American Society for Testing and Materials), 100 Barr Harbor Drive, West Conshohocken, PA, USA
14411	06	Ceramic tiles - Definitions, classification, characteristics and marking. [national adoptions from EN: BS DIN I.S. NBN NEN NF NS ÖNORM PN SN SS UNE UNI]
14774-1	09	Solid biofuels - Determination of moisture content - Oven dry method - Part 1: Total moisture - Reference method. [national adoptions from EN: BS DIN I.S. NBN NEN NF NS ÖNORM PN SN SS UNI]
14774-2	09	Solid biofuels - Determination of moisture content - Oven dry method - Part 2: Total moisture - Simplified method. [national adoptions from EN: BS DIN I.S. NBN NEN NF NS ÖNORM PN SN SS UNI]
14774-3	09	Solid biofuels - Determination of moisture content - Oven dry method - Part 3: Moisture in general analysis sample. [national adoptions from EN: BS DIN I.S. NBN NEN NF NS ÖNORM PN SN SS UNI]
20535	94	≡ ISO 535: Paper and boards - Determination of water absorptiveness - Cobb method. [national adoptions from EN: BS DIN I.S. NBN NF ÖNORM PN SN SS UNE UNI]
29865	93	≡ ISO 9865: Textiles - Determination of water repellency of fabrics by the Bundesmann rain-shower test. [national adoptions from EN: BS DIN I.S. NBN NF ÖNORM PN SN SS UNE UNI]
60749-7	02	Semiconductor devices - Mechanical and climatic test methods - Part 7: Internal moisture content measurement and the analysis of other residual gases. [national adoptions from EN: BS CEI DIN IEC I.S. NBN NEN NF ÖNORM PN SS UNE]
60749-7	06	Semiconductor devices - Mechanical and climatic test methods - Part 39: Measurement of moisture diffusivity and water solubility in organic materials used for semiconductor components. [national adoptions from EN: BS CEI DIN IEC I.S. NBN NEN NF ÖNORM PN SS]
60811-1-3	95	≡ IEC 60811-1-3 ≡ VDE 0473-811-1-3: Insulating and sheathing materials of electric cables - Common test methods - Part 1: General application; Section 3: Methods for determining the density - Water absorption tests - Shrinkage test. [national adoptions from EN: BS CEI DIN I.S. NBN NEN NF ÖNORM PN SN SS UNE]

Table 4. (Continued)

ASTM		ASTM International (former American Society for Testing and Materials), 100 Barr Harbor Drive, West Conshohocken, PA, USA
60893-1	04	≡ IEC 60893-1 ≡ VDE 0318-1: Insulating materials - Industrial rigid laminated sheets based on thermosetting resins for electrical purposes - Part 1: Definitions, designations and general requirements. [national adoptions from EN: BS CEI DIN I.S. NBN NEN NF ÖNORM PN SN SS UNE]
EN ISO		**Comité Européenne de Normalisation - CEN, Rue de Stassart 36, B-1050 Bruxelles, Belgium**
62		≡ ISO 62: Plastics - Determination of water absorption. [national adoptions from EN ISO: BS DIN I.S. NBN NEN NF NS ÖNORM PN SN SS UNE UNI]
287	09	≡ ISO 287: Paper and board - Determination of moisture content of a lot - Oven-drying method. [national adoptions from EN ISO: BS DIN I.S. NBN NEN NF NS ÖNORM PN SN SS UNE UNI]
291	08	≡ ISO 291: Plastics - Standard atmospheres for conditioning and testing. [national adoptions from EN ISO: BS DIN I.S. NBN NEN NF NS ÖNORM PN SN SS UNE UNI]
585	99	≡ ISO 585: Plastics - Unplasticized cellulose acetate - Determination of moisture content. [national adoptions from EN ISO: BS DIN I.S. NBN NEN NF NS ÖNORM PN SN SS UNE UNI]
638	08	≡ ISO 638: Paper, board and pulps - Determination of dry matter content - Oven-drying method. [national adoptions from EN ISO: BS DIN I.S. NBN NEN NF NS PN SN SS UNE UNI]
665	00	≡ ISO 665: Oilseeds - Determination of moisture and volatile matter content. [national adoptions from EN ISO: BS DIN I.S. NBN NEN NF NS ÖNORM PN SN SS UNE UNI]
712	09	≡ ISO 712: Cereals and cereal products - Determination of moisture content - Reference method. [national adoptions from EN ISO: BS DIN I.S. NBN NEN NF NS ÖNORM PN SN SS UNE UNI]
1517	95	≡ ISO 1517 (superseded by EN ISO 9117-3): Paints and varnishes - Surface-drying test - Ballotini method. [national adoptions from EN ISO: DIN NBN NF NS ÖNORM SN]

ASTM		ASTM International (former American Society for Testing and Materials), 100 Barr Harbor Drive, West Conshohocken, PA, USA
1666	97	≡ ISO 1666: Starch - Determination of moisture content - Oven-drying method. [national adoptions from EN ISO: BS DIN I.S. NBN NEN NF NS ÖNORM PN SN SS UNE UNI]
1741	94	≡ ISO 1741: Dextrose - Determination of loss in mass on drying - Vacuum oven method. [national adoptions from EN ISO: BS DIN I.S. NBN NEN NF NS ÖNORM PN SN SS UNE UNI]
2417	02	≡ ISO 2417: Leather - Physical and mechanical tests - Determination of the static absorption of water. [national adoptions from EN ISO: BS DIN I.S. NBN NEN NF NS ÖNORM PN SN SS UNE UNI]
3251		≡ BS 3900-B18(2003) ≡ ISO 3251: Paints, varnishes and plastics - Determination of non-volatile-matter content. [national adoptions from EN ISO: BS DIN I.S. NBN NEN NF NS ÖNORM PN SN SS UNE UNI]
3344	97	≡ ISO 3344: Reinforcement products - Determination of moisture content. [national adoptions from EN ISO: BS DIN I.S. NBN NEN NF NS ÖNORM PN SN SS UNE UNI]
3727-1	01	≡ ISO 3727-1: Butter - Determination of moisture, non-fat solids and fat contents - Part 1: Determination of moisture content (reference method). [national adoptions from EN ISO: BS DIN I.S. NBN NEN NF NS ÖNORM PN SN SS UNE UNI]
5404	02	≡ ISO 5404: Leather - Physical and mechanical tests - Determination of the water resistance of heavy leathers. [national adoptions from EN ISO: BS DIN I.S. NBN NEN NF NS ÖNORM PN SN SS UNE UNI]
5537	04	≡ ISO 5537: Dried milk - Determination of moisture content (reference method). [national adoptions from EN ISO: BS DIN I.S. NBN NEN NF NS ÖNORM PN SN SS UNE UNI]
6330	00	≡ ISO 6330: Textiles - Domestic washing and drying procedures for textile testing. [national adoptions from EN ISO: BS DIN I.S. NBN NEN NF NS ÖNORM PN SN SS UNE UNI]
6540	10	≡ ISO 6540: Maize - Determination of moisture content (on milled grains and on whole grains). [national adoption from EN ISO: I.S.]

Table 4. (Continued)

ASTM		ASTM International (former American Society for Testing and Materials), 100 Barr Harbor Drive, West Conshohocken, PA, USA
9117-1	09	≡ ISO 9117-1: Paints and varnishes - Drying tests - Part 1: Determination of through-dry state and through-dry time. [national adoptions from EN ISO: BS DIN I.S. NBN NEN NF NS PN SN SS UNE UNI]
9117-2	10	Paints and varnishes - Drying tests - Part 2: Pressure test for stackability. [national adoptions from EN ISO: BS I.S. NEN PN SS UNI]
9117-3	10	≡ ISO 9665: Paints and varnishes - Drying tests - Part 3: Surface-drying test using ballotini. [national adoptions from EN ISO: BS I.S. NEN NF PN SS UNI]
9665	00	Adhesives - Animal glues - Methods for sampling and testing. [national adoptions from EN ISO: BS DIN I.S. NBN NEN NF NS ÖNORM PN SN SS UNE UNI]
10545-3	97	≡ ISO10545-3: Ceramic tiles - Part 3: Determination of water absorption, apparent porosity, apparent relative density and bulk density. [national adoptions from EN ISO: BS DIN I.S. NBN NEN NF NS ÖNORM PN SN SS UNE UNI]
10565	98	≡ ISO 10565: ≡ BS 4289-11(1998): Oilseeds - Simultaneous determination of oil and water contents - Method using pulsed nuclear magnetic resonance spectrometry. [national adoptions from EN ISO: BS DIN I.S. NBN NEN NF NS ÖNORM PN SN SS UNE UNI]
11125-7	97	≡ ISO 11125-7: Preparation of steel substrates before application of paints and related products - Test methods for metallic blast-cleaning abrasives - Part 7: Determination of moisture. [national adoptions from EN ISO: BS DIN I.S. NBN NEN NF NS ÖNORM PN SN SS UNE UNI]
11127-5	97	≡ ISO 11127-5: Preparation of steel substrates before application of paints and related products - Test methods for non-metallic blast-cleaning abrasives - Part 5: Determination of moisture. [national adoptions from EN ISO: BS DIN I.S. NBN NEN NF NS ÖNORM PN SN SS UNE UNI]
12570	00	≡ ISO 12570: Hygrothermal performance of building materials and products - Determination of moisture content by drying at elevated temperature. [national adoptions from EN ISO: BS DIN I.S. NBN NEN NF NS ÖNORM PN SN SS UNE UNI]

ASTM		ASTM International (former American Society for Testing and Materials), 100 Barr Harbor Drive, West Conshohocken, PA, USA
12571	00	≡ ISO 12571: Hygrothermal performance of building materials and products - determination of hygroscopic sorption properties. [national adoptions from EN ISO: BS DIN I.S. NBN NEN NF NS ÖNORM PN SN SS UNE UNI]
13758	96	≡ ISO 13758: Methods of test for petroleum and its products - Liquefied petroleum gases - Assessment of the dryness of propane - Valve freeze method. [national adoptions from EN ISO: BS DIN I.S. NBN NEN NF NS ÖNORM PN SN SS UNE UNI]
15148	02	≡ ISO 15148: Hygrothermal performance of building materials and products - Determination of water absorption coefficient by partial immersion. [national adoptions from EN ISO: BS DIN I.S. NBN NEN NF NS ÖNORM PN SN SS UNE UNI]
60811-1-3	95	≡ IEC 60811-1-3 ≡ VDE 0473-811-1-3: Insulating and sheathing materials of electric cables - Common test methods - Part 1: General application; Section 3: Methods for determining the density - Water absorption tests - Shrinkage test.
60893-1	04	≡ IEC 60893-1 ≡ VDE 0318-1: Insulating materials - Industrial rigid laminated sheets based on thermosetting resins for electrical purposes - Part 1: Definitions, designations and general requirements. [national adoptions from EN ISO: BS CEI DIN I.S. NBN NEN NF ÖNORM PN SN SS UNE]

prEN		Comité Européenne de Normalisation - CEN, Rue de Stassart 36, B-1050 Bruxelles, Belgium
3615	99	≡ DIN EN 3615 ≡ NFL 17 457: Aerospace series - Fibre reinforced plastics - Determination of the conditions of exposure to humid atmosphere and of moisture absorption.
15414-3	09	Solid recovered fuels - Determination of moisture content using the oven dry method - Part 3: Moisture in general analysis sample.

GOST		Committee of the Russian Federation for Standardization, Metrology and Certification, Leninsky Prospekt 9, Moskva 117049, Russia
2409	95	Refractories - Method for determination of bulk density apparent and true porosity, water absorption.
12730.0	78	Concretes - General requirements to methods of determination of density, porosity, moisture content, water absorptions and water tightness.

Table 4. (Continued)

ASTM		ASTM International (former American Society for Testing and Materials), 100 Barr Harbor Drive, West Conshohocken, PA, USA
IEC		International Electrotechnical Commission, 3 rue de Varembé, P.O. Box 131, CH-1211 Geneva 20, Switzerland
60811-1-3	01	≡ EN 60811-1-3 ≡ GOST R IEC 60811-1-3 ≡ JIS C3660-1-3 ≡ NFC 32 026 Common test methods for insulating and sheathing materials of electric cables - Methods for general application - Methods for determining the density - Water absorption tests - Shrinkage test.
ISO		International Organization for Standardization, Central Secretariat, 1 rue de Varembé, BP 56, CH- 1211 Geneva 20, Switzerland
62	08	Plastics - Determination of water absorption. [direct adoption from ISO: EN (national adoptions from EN see under EN ISO 62)]
287	09	≡ JIS P8127: Paper and board - Determination of moisture content of a lot - Oven-drying method. [direct adoption from ISO: EN (national adoptions from EN see under EN ISO 287)]
291	08	Plastics - Standard atmospheres for conditioning and testing. [direct adoption from ISO: EN (national adoptions from EN see under EN ISO 291)]
310	92	Manganese ores and concentrates - Determination of hygroscopic moisture content in analytical samples - Gravimetric method. [direct adoption from ISO: PN]
535	91	≡ EN 20535 ≡ JIS P8140: Paper and board - Determination of water absorption - Cobb method. [direct adoptions from ISO: NBR NM NS]
585	90	Plastics - Non plasticized cellulose acetate - Determination of moisture content. [direct adoption from ISO: EN (national adoptions from EN see under EN ISO 585)]
589	08	≡ SANS 589: Hard coal - Determination of total moisture. [direct adoption from ISO: NEN]
638	08	≡ JIS P8203: Paper, board and pulps - Determination of dry matter content - Oven-drying method. [direct adoption from ISO: EN (national adoptions from EN see under EN ISO 638)]
665	00	≡ EN ISO 665: Oilseeds - Determination of moisture and volatile matter content. [direct adoption from ISO: EN (national adoptions from EN see under EN ISO 665)]

ASTM		ASTM International (former American Society for Testing and Materials), 100 Barr Harbor Drive, West Conshohocken, PA, USA
672	78	≡ BS 1715-2.6(1989): Soaps - Determination of moisture and volatile matter content - Oven method. [direct adoptions from ISO: NBN NEN PN UNE]
711	85	≡ DIN 10350 ≡ NFV 03 706 ≡ PN 93/A-74012 ≡ UNE 34068-2: Cereals and cereal products - Determination of moisture content (Basic reference method). [direct adoptions from ISO: NF UNI]
712	09	Cereals and cereal products - Determination of moisture content - Reference method. [direct adoption from ISO: EN (national adoptions from EN see under EN ISO 712)]
771	77	≡ DIN 10268: Oilseed residues - Determination of moisture and volatile matter content. [direct adoption from ISO: PN]
802	76	≡ NBN T 03 401 ≡ SIS 11 77 90: Aluminium oxide primarily used for the production of aluminium - Preparation and storage of test samples. [direct adoption from ISO: SS]
939	80	≡ UNE 34096-7 ≡ : NFV 03 402: Spices and condiments - Determination of moisture content - Entrainment method. [direct adoptions from ISO: DIN PN]
1026	82	Fruit and vegetable products - Determination of dry matter content by drying under reduced pressure and of water content by azeotropic distillation. [direct adoption from ISO: PN]
1442	97	≡ ÖNORM N 4004 ≡ UNE 34552-2: Meat and meat products - determination of moisture content (Reference method). [national adoptions from ISO: NEN PN UNI]
1609	96	≡ EN 1609: Thermal insulating products for building applications - Determination of short term water absorption by partial immersion.
1666	96	Starch - Determination of moisture content - Oven-drying method. [direct adoption from ISO: EN (national adoptions from EN see under EN ISO 1666)]
1741	94	≡ NFV 03 621: Dextrose - Determination of loss in mass on drying - Vacuum oven method. [direct adoption from ISO: EN (national EN adoptions see under EN ISO 1741)]
2066	04	Resilient floor coverings - Determination of moisture content of agglomerated composition cork.

Table 4. (Continued)

ASTM		ASTM International (former American Society for Testing and Materials), 100 Barr Harbor Drive, West Conshohocken, PA, USA
2190	98	≡ UNE 56917: Cork - granulated cork - determination of moisture content. [direct adoption from ISO: NF]
2291	80	Cocoa beans - Determination of moisture content (Routine method).
2386	98	Corkwood in planks, virgin cork, ramassage, gleanings, corkwood refuse and corkwaste - determination of moisture content. [direct adoption from ISO: NF]
2417	02	≡ UNE 59006: Leather - Physical and mechanical tests - Determination of the static absorption of water. [direct adoption from ISO: SANS, EN (national adoptions from EN see under EN ISO 1741)]
2462	73	≡ NBN T 03 152: Sodium chlorate for industrial use - Determination of moisture content - Gravimetric method.
2483	73	≡ NBN T 03 165 ≡ UNE 34203: Sodium chloride for industrial use - Determination of the loss of mass at 110 °C.
2596	06	Iron ores - determination of hygroscopic moisture in analytical samples - Gravimetric, Karl Fischer and mass-loss methods. [direct adoptions from ISO: BN NEN SS]
2696	76	Fibre building boards - Hard and medium boards for general purposes - Quality specifications - Water absorption and swelling in thickness.
2832	73	≡ NBN T 03 462 ≡ NFT 20 151: Sodium fluoride for industrial use - Determination of moisture content. [direct adoption from ISO: PN]
2896	87	Rigid cellular plastics - Determination of water absorption. [direct adoptions from ISO: NEN SS]
3082	09	Iron ores - Sampling and sample preparation procedures. [direct adoptions from ISO: BS NEN]
3087	98	Iron ores - Determination of the moisture content of a lot. [direct adoptions from ISO: BS NBR PN UNI]
3130	75	≡ DIN 52183: Wood - Determination of moisture content for physical and mechanical tests. [direct adoptions from ISO: NBN ÖNORM UNI]
3251	08	≡ BS 3900-B18(2003): Paints, varnishes and plastics - Determination of non-volatile-matter content. [direct adoption from ISO: EN (national adoptions from EN see under EN ISO 3251]
3344	97	Reinforcement products - Determination of moisture content. [direct adoption from ISO: EN (national adoptions from EN see under EN ISO 3344]

ASTM		ASTM International (former American Society for Testing and Materials), 100 Barr Harbor Drive, West Conshohocken, PA, USA
3393	76	≡ NBN T 03 444 ≡ NFT 20 161: Cryolite, natural and artificial, and aluminium fluoride for industrial use determination of moisture content - Gravimetric method. [direct adoption from ISO: SS]
3727-1	01	Butter - Determination of moisture, non-fat solids and Fat contents - Part 1: Determination of moisture content (Reference method). [direct adoption from ISO: EN (national EN adoptions see under EN ISO 3727-1)]
4299	89	Manganese ores - Determination of moisture content.
4470	81	Sawn timber - determination of the average moisture content of a lot. [direct adoption from ISO: ÖNORM]
5068-1	07	Brown coals and lignites - Determination of moisture content - Part 1: Indirect gravimetric method for total moisture. [direct adoptions from ISO: BS NEN]
5068-2	07	Brown coals and lignites - Determination of moisture content - Part 2: Indirect gravimetric method for moisture in the analysis sample. [direct adoptions from ISO: BS NEN]
5069-1	83	Brown coals and lignites - Principles of sampling - Part 1: Sampling for determination of moisture content and for general analysis. [direct adoption from ISO: PN]
5069-2		Brown coals and lignites - Principles of sampling - Part 2: Sample preparation for determination of moisture content and for general analysis. [direct adoption from ISO: PN]
5404	02	≡ UNE 59007: Leather - Physical and mechanical tests - Determination of the water resistance of heavy leathers. [direct adoption from ISO: EN (national adoptions from EN see under EN ISO 5404]
5530-1	97	≡ BIS IS 12516-1: Wheat flour - Physical characteristics of doughs - Part 1: Determination of water absorption and rheological properties using a farinograph. [direct adoption from ISO: NEN NF PN]
5530-3	88	Wheat flour - Physical characteristics of doughs - Part 3: Determination of water absorption and rheological properties using a valorigraph.
5537	04	≡ BIS IS 11623 ≡ SANS 5537: Dried milk - Determination of moisture content (Reference method). [direct adoption from ISO: BS DIN I.S. NBN NEN NF NS ÖNORM PN SN SS UNE UNI]
5550	06	Caseins and caseinates - Determination of moisture content (reference method). [direct adoptions from ISO: BS NEN NF]

Table 4. (Continued)

ASTM		ASTM International (former American Society for Testing and Materials), 100 Barr Harbor Drive, West Conshohocken, PA, USA
5635	78	≡ PN 92/P-50158 ≡ UNE 57049: Paper - Measurement of dimensional change after immersion in water.
5637	89	≡ AS/NZS 1301.442S ≡ PN 92/P-50159 ≡ UNE 57112: Paper and board - Determination of water absorption and increase in thickness after immersion in water.
6129	81	Chromium ores - Determination of hygroscopic moisture content in analytical samples - Gravimetric method.
6330	00	≡ AS 2001.5.4 ≡ BIS IS 15370 ≡ EN 26330: Textiles - Domestic washing and drying procedures for textile testing. [direct adoptions from ISO: EN GOST-R (national adoptions from EN see under EN ISO 6330)]
6540	80	Maize - Determination of moisture content (on milled grains and on whole grains). [direct adoptions from ISO: DIN EN I.S. PN (national adoptions from EN see under EN ISO 6540)]
6783	82	≡ DIN 52102 ≡ SS 13 21 25 ≡ UNE 83134: Coarse aggregates for concrete - Determination of particle density and water absorption hydrostatic balance method.
7033	87	Fine and coarse aggregates for concrete - Determination of the particle mass-per-volume and water absorption - Pycnometer method.
7513	90	≡ BIS IS 13859: Instant tea in solid form - Determination of moisture content (loss in mass at 103 °C) [direct adoptions from ISO: NF PN]
7700-1	08	Food products - checking the performance of moisture meters in use - Part 1: Moisture meters for cereals. [direct adoptions from ISO: BS NF]
7700-2	87	Check of the calibration of moisture meters - Part 2: Moisture meters for oilseeds.
8361-2	91	Thermoplastics pipes and fittings - Water absorption - Test conditions for unplasticized poly vinyl chloride (PVC-U) pipes and fittings. [direct adoptions from ISO: NEN PN]
8361-3	91	≡ DIN 16890: Thermoplastics pipes and fittings - Water absorption - Test conditions for acrylonitrile butadiene styrene (ABS) pipes and fittings. [direct adoption from ISO: PN]
8375	09	Timber structures - Glued laminated timber – Test methods for determination of physical and mechanical properties.
8557	85	Aluminium ores - Determination of hygroscopic moisture in analytical samples - Gravimetric method.

ASTM		ASTM International (former American Society for Testing and Materials), 100 Barr Harbor Drive, West Conshohocken, PA, USA
8756	94	Air quality - Handling of temperature, pressure and humidity data. [direct adoptions from ISO: DIN GOST R PN SS]
8787	86	≡ JIS P8141 ≡ UNE 57044: Paper and board - Determination of capillary rise - Klemm method. [direct adoptions from ISO: DIN EN NBR NM NEN NF PN UNI]
8851-1	04	Butter - Determination of moisture, non-fat solids and fat contents (routine methods) - Part 1: Determination of moisture content. [direct adoptions from ISO: NEN NF]
9033	89	≡ AS 2932.2: Aluminium ores - Determination of the moisture content of bulk material.
9117-1	09	≡ EN ISO 9117-1: Paints and varnishes - Drying tests - Part 1: Determination of through-dry state and through-dry time. [direct adoption from ISO: EN (national adoptions from EN see under EN ISO 9117-1]
9117-2	10	Paints and varnishes - Drying tests - Part 2: Pressure test for stackability. [direct adoption from ISO: EN (national adoptions from EN see under EN ISO 9117-2]
9117-3	10	Paints and varnishes - Drying tests - Part 3: Surface-drying test using ballotini. [direct adoption from ISO: EN (national adoptions from EN see under EN ISO 9117-3]
9665	98	≡ BS 647(1993): Adhesives - Animal glues - Methods for sampling and testing. [direct adoption from ISO: EN (national adoptions from EN see under EN ISO 9665]
9865	93	≡ EN 29865: Textiles - Determination of water repellency of fabrics by the Bundesmann rain-shower test. [direct adoptions from ISO: NEN NS]
10251	06	≡ AS 2863: Copper, lead, zinc and nickel concentrates - Determination of mass loss of bulk material on drying. [direct adoption from ISO: BS]
10545-3	97	≡ SAC GB/T 3810.3 ≡ SIA 248.078: Ceramic tiles - Part 3: Determination of water absorption, apparent porosity, apparent relative density and bulk density. [direct adoption from ISO: EN (national adoptions from EN see under EN ISO 10545-3]

Table 4. (Continued)

ASTM		ASTM International (former American Society for Testing and Materials), 100 Barr Harbor Drive, West Conshohocken, PA, USA
10565	98	Oilseeds - Simultaneous determination of oil and water contents - Method using pulsed nuclear magnetic resonance spectrometry. [direct adoption from ISO: EN (national adoptions from EN see under EN ISO 10565]
11093-3	94	≡ UNE 57163-3: Paper and board - Testing of cores - Part 3: Determination of moisture content using the oven drying method. [direct adoptions from ISO: BS DIN NF]
11125-7	93	≡ JIS Z0311 ≡ NFT 35 507-7: Preparation of steel substrates before application of paints and related products - Test methods for metallic blast-cleaning abrasives - Part 7: Determination of moisture. [direct adoption from ISO: EN (national adoptions from EN see under EN ISO 11125-7]
11127-5	93	≡ JIS Z0312: Preparation of steel substrates before application of paints and related products - Test methods for non-metallic blast-cleaning abrasives - Part 5: Determination of moisture. [direct adoption from ISO: EN (national adoptions from EN see under EN ISO 9117-3]
11294	94	Roasted ground coffee - Determination of moisture content - Method by determination of loss in mass at 103 °C (Routine method). [direct adoptions from ISO: NF PN]
11461	01	Soil quality - Determination of soil water content as a volume fraction using coring sleeves - Gravimetric method. [direct adoptions from ISO: BS DIN NF PN SS UNE]
11464	06	Soil quality - Pretreatment of samples for physico-chemical analyses. [direct adoptions from ISO: BS DIN NF SS]
11465	93	≡ BIS IS 15106: Soil quality - Determination of dry matter and water content on a mass basis - Gravimetric method. [direct adoptions from ISO: DIN NEN NF PN SS]
11520-2	01	Agricultural grain driers - Determination of drying performance - Part 2: Additional procedures and cropspecific requirements. [direct adoptions from ISO: BS NEN SS]
11722	99	≡ DIN 51718 ≡ JIS M8812 ≡ SANS 11722: Solid mineral fuels - Hard coal - Determination of moisture in the general analysis test sample by drying in nitrogen. [direct adoptions from ISO: NEN PN]
11817	94	≡ DIN 10772-1: Roasted ground coffee - Determination of moisture content - Karl Fischer method (Reference method). [direct adoptions from ISO: NF PN]

ASTM		ASTM International (former American Society for Testing and Materials), 100 Barr Harbor Drive, West Conshohocken, PA, USA
12570	00	≡ SIA 180.214: Hygrothermal performance of building materials and products - Determination of moisture content by drying at elevated temperature. [direct adoption from ISO: EN (national adoptions from EN see under EN ISO 12570]
12571	00	Hygrothermal performance of building materials and products - Determination of hygroscopic sorption properties. [direct adoption from ISO: EN (national adoptions from EN see under EN ISO 12571]
12572	01	Hygrothermal performance of building materials and products – determination of water vapour transmission properties
12743	06	≡ AS 2862.1: Copper, lead, zinc and nickel concentrates – Sampling procedures for determination of metal and moisture content. [direct adoption from ISO: BS]
13758	96	≡ IP 395: Methods of test for petroleum and its products - Liquefied petroleum gases - Assessment of the dryness of propane - Valve freeze method. [direct adoption from ISO: EN (national adoptions from EN see under EN ISO 13758]
13765-6	04	Refractory mortars - Part 6: Determination of moisture content of ready-mixed mortars. [direct adoptions from ISO: BS NEN]
15148	02	Hygrothermal performance of building materials and products - Determination of water absorption coefficient by partial immersion. [direct adoption from ISO: EN (national adoptions from EN see under EN ISO 15148]
15968	00	Direct reduced iron - Determination of apparent density and water absorption of hot briquetted iron (HBI).
16586	03	Soil quality - Determination of soil water content as a volume fraction on the basis of known dry bulk density - Gravimetric method. [direct adoptions from ISO: DIN NF PN UNE]
17190-4	01	Urine-absorbing aids for incontinence - Test methods for characterizing polymer-based absorbent materials -Part 4: Determination of moisture content by mass loss upon heating. [direct adoption from ISO: NEN]
18775	08	Veneers - Terms and definitions, determination of physical characteristics and tolerances. [direct adoption from ISO: NF]
20983	08	Instant coffee - determination of moisture content -Karl Fischer method (Reference method). [direct adoptions from ISO: BS NEN]

Table 4. (Continued)

ASTM		ASTM International (former American Society for Testing and Materials), 100 Barr Harbor Drive, West Conshohocken, PA, USA
23391	06	Dried rosehips - Specification and test methods.
23394	06	Dried oleaster - Specification and test methods.
24557	09	Pulses - Determination of moisture content - Air-oven method. [direct adoptions from ISO: BS NEN]

JIS		Japanese Standards Association (JSA) 1-24 Akasaka 4-Chome Minato-ku, Tokyo 107, Japan
A 1111	07	Method of test for surface moisture in fine aggregate.
A 1125	07	Methods of test for moisture content of aggregate and surface moisture in aggregate by drying.
A 1437	92	Test methods for moisture resistance of boards for interior system of building.
A 1470-1	08	Determination of water vapour adsorption/desorption properties for building materials - Part 1: Response to humidity variation.
A 1470-2	08	Determination of water vapour adsorption/desorption properties for building materials - Part 2: Response to temperature variation.
A 1475	04	Method of test for hygroscopic sorption properties of building material.
A 1476	06	Measuring method for moisture content of building materials by drying at elevated temperature.
A 1802	09	Methods of test for production control of concrete - Method of test for surface moisture in fine aggregate by centrifugal force.
A 1803	09	Methods of test for production control of concrete - Method of test for surface moisture in coarse aggregate.
A 6930	97	Plastic films vapour barrier for residential use.
B 8392-3	01	Compressed air - Part 3: Test methods for measurement of humidity.
C 9730-2-13	10	Automatic electrical controls for household and similar use - Part 2-13: Particular requirements for humidity sensing controls.
C 60068-2-38	88	Basic environmental testing procedures - Part 2: Tests, Test Z/AD: Composite temperature/humidity cyclic test
C 60068-3-6	08	Environmental testing - Part 3-6: Supporting documentation and guidance - Confirmation of the performance of temperature/humidity chambers.
C 60721-2-1	95	Classification of environmental conditions - Part 2: Environmental conditions appearing in nature temperature and humidity.
C 60721-3-4	97	Classification of environmental conditions - Part 3: Classification of groups of environmental parameters and their severities - Section 4: Stationary use at non-weatherprotected locations.
C 61300-2-48	10	Fiber optic interconnecting devices and passive components - Basic test and measurement procedures - Part 2-48: Tests - Temperature - humidity cycling
D 0203	94	Method of moisture, rain and spray test for automobile parts.

ASTM		ASTM International (former American Society for Testing and Materials), 100 Barr Harbor Drive, West Conshohocken, PA, USA
D 4414-1	98	Automotive parts - Brake linings and disc brake pads - Part 1: Test procedure of seizure to ferrous mating surface due to corrosion (moisture adsorption method).
E 3017	07	Parts for railway signalling - Waterproof test method.
E 4034	94	Railway rolling stock parts - Test methods of resistibility for moisture and rainfall.
G 2403	03	Methods for sampling, preparation and determination of moisture content of aluminium dross.
K 0225	02	Testing methods for determination of trace components in diluent gas and zero gas.
K 1468-1	99	Acid-grade fluorspar - Method for chemical analysis - Part 1: Determination of moisture content of a lot.
K 2301	08	Fuel gases and natural gas - Methods for chemical analysis and testing.
K 6857	73	Testing methods for resistance of adhesive bonds to water or moisture.
K 5600-3-2	99	≡ ISO 1517: Testing methods for paints - Part 3: Film formability - Section 2: Surface-drying test (Ballotini method).
K 5600-3-3	99	≡ ISO 9117: Testing methods for paints - Part 3: Film formability - Section 3: Through-dry.
K 5600-7-2	99	Testing methods for paints - Part 7: Long-period performance of film - Section 2: Resistance to humidity (continuous condensation).
K 5600-7-3	99	Testing methods for paints - Part 7: Long-period performance of film - Section 3: Resistance to humidity (intermittent condensation).
K 5600-7-4	99	Testing methods for paints - Part 7: Long-period performance of film - Section 4: Humidity and cool-heat cycling test.
K 5600-7-9	06	Testing methods for paints - Part 7: Determination of resistance to cyclic corrosion conditions - Section 9: Salt fog/dry/humidity
K 7129	08	Plastics - Film and sheeting - Determination of water vapour transmission rate - Instrumental method.
K 7225	05	Rigid cellular plastics - Determination of water vapour transmission properties.
L 1915	98	Test methods of humidity transfer characteristics of futon.
M 8082	99	Method of sampling and methods of determination of copper content and moisture of copper scraps for smelting.
M 8100	92	Particulate materials - General rules for methods of sampling.
M 8101	88	Methods for sampling, preparation and determination of moisture content of non-ferrous metal bearing ores.
M 8102	93	Methods for sampling and determination of moisture content of blister copper.
M 8108	08	Chromium, manganese and ferruginous manganese ores - Sampling, sample preparation and determination of moisture content and size distribution.

Table 4. (Continued)

ASTM		ASTM International (former American Society for Testing and Materials), 100 Barr Harbor Drive, West Conshohocken, PA, USA
M 8109	96	Method for sampling and method of determination of moisture content of garnierite nickel ores.
M 8705	06	Iron ores - Determination of the moisture content of a lot.
M 8803	93	Methods for test of equilibrium moisture of coal at 96 to 97 % relative humidity and 30 °C
M 8811	00	Methods for sampling and determination of total moisture and adherent moisture of coal and coke.
M 8820	00	Coal and coke - Determination of total moisture content of a lot.
P 8127	10	Paper and board - Determination of moisture content of a lot - Oven-drying method.
P 8153	06	Paper and board - Measurement of hygroexpansivity - Part 1: Hygroexpansivity up to a maximum relative humidity of 68 %.
P 8154-3	08	Paper and board - Accelerated ageing - Part 3: Moist heat treatment at 80 °C and 65 % relative humidity.
P 8202	98	Method of testing water content of lot of paper pulp.
P 8203	10	≡ ISO 638: Paper, board and pulps - Determination of dry matter content - Oven-drying method.
R 1703-1	07	Fine ceramics (advanced ceramics, advanced technical ceramics) - Test method for self-cleaning performance of photocatalytic materials - Part 1: Measurement of water contact angle.
R 3913	00	Reinforcement products - Determination of moisture content.
W 0814	02	Aircraft equipment - Environmental and operating conditions for airborne equipment - Humidity, temperature and pressure tests.
Z 0208	76	Testing methods for determination of the water vapour transmission rate of moisture-proof packaging materials (dish method).
Z 0222	59	Method of permeability test for moisture proof packing case.
Z 0301	89	Method of moisture-proof packaging.
Z 7302-3	99	Densified refuse derived fuel - Part 3: Test method for moisture.
Z 8806	01	Humidity - Measurement methods.
JLPGA		Japanese Liquefied Petroleum Gas Association
S-03	81	LP Gas evaporation residue test method (Mass method)
LN		Luftfahrt-Norm, Deutsches Institut für Normung e.V., Burggrafenstraße 6, D-10772 Berlin, Germany
29820	77	Polydiallylphthalate (PDAP) moulding compounds with glass-fibre filler; Technical specification.
65336	73	Sheets, tubes and rods of laminates on phenol-formaldehyde resin base - Technical specification.

ASTM		ASTM International (former American Society for Testing and Materials), 100 Barr Harbor Drive, West Conshohocken, PA, USA
SFS		Suomen Standardisoimisliitlo r.y. NT Build 361, Nordtest, Espoo
	91	Nordtest method, Concrete, Hardened: Water-cement ratio,
SIA		Schweizerischer Ingenieur- und Architektenverein, Selnaustrasse 16, CH-8001 Zürich
262.175	05	≡ SN EN 480-5: Admixtures for concrete, mortar and grout - Test methods - Part 5: Determination of capillary absorption.
164.504	97	≡ SN EN 844-4: Round and sawn timber - Terminology - Part 4: Terms relating to moisture content.
266.016	05	≡ SN EN 771-6: Specification for masonry units - Part 6: Natural stone masonry units.
177.107	98	≡ SN EN 772-7: Methods of test for masonry units - Part 7: Determination of water absorption of clay masonry damp proof course units by boiling in water.
177.111	00	≡ SN EN 772-11 Methods of test for masonry units - Part 11: Determination of water absorption of aggregate concrete, manufactured stone and natural stone masonry units due to capillary action and the initial rate of water absorption of clay masonry units.
177.160	99	≡ SN EN 1015-10 Methods of test for mortar for masonry - Part 10: Determination of dry bulk density of hardened mortar.
180.214	00	≡ SN EN ISO 12570 ≡ ISO 12570: Hygrothermal performance of building materials and products - Determination of moisture content by drying at elevated temperature.
246.201	99	≡ SN EN 1925: Natural stone test methods - Determination of water absorption coefficient by capillarity.
246.211	08	≡ SN EN 13755: Natural stone test methods - Determination of water absorption at atmospheric pressure.
248.078	97	≡ SN EN ISO 10545-3: Ceramic tiles - Part 3: Determination of water absorption, apparent porosity, apparent relative density and bulk density.
253.049	98	≡ SN EN 12105: Resilient floor coverings - Determination of moisture content of agglomerated composition cork.
279.074	96	≡ SN EN 1609: Thermal insulating products for building applications - Determination of short term water absorption by partial immersion.
279.077	97	≡ SN EN 12087: Thermal insulating products for building applications - Determination of long term water absorption by immersion.
279.078	97	≡ SN EN 12088: Thermal insulating products for building applications - Determination of long term water absorption by diffusion.

Table 4. (Continued)

ASTM		ASTM International (former American Society for Testing and Materials), 100 Barr Harbor Drive, West Conshohocken, PA, USA
VDE		Verband der Elektrotechnik Elektronik Informationstechnik e.V., Stresemannallee 15, D-60596 Frankfurt am Main
0291-2	79	≡ DIN 57291-2: Compounds for use in cable accessories - Casting resinous compounds before cure and in the cured state.
0318-1	76	≡ EN 60893-1: Insulating materials - Industrial rigid laminated sheets based on thermosetting resins for electrical purposes - Part 1: Definitions, designations and general requirements.
0318-2	76	≡ EN 60893-2: Insulating materials - Industrial rigid laminated sheets based on thermosetting resins for electrical purposes - Part 2: Methods of test.
VDMA		Verband Deutscher Maschinen- und Anlagenbau e.V., Postfach 71 08 64, D-60498 Frankfurt am Main
24351	99	Drying technology - General terms.

Table 5. List of national identifiers for standards and the assigned standardization bodies

AS	Standards Australia, GPO Box 476, Sydney NSW 2001, Australia
ASTM	ASTM International (former American Society for Testing and Materials), 100 Barr Harbor Drive, West Conshohocken, PA, USA
BS	BSI British Standards Institution, 389 Chiswick High Road, London, W4 4AL, U.K.
BVL	Bundesamt für Verbraucherschutz und Lebensmittelsicherheit, Bundesallee 50, 38116 Braunschweig, Germany
CEI	Comitato Elettrotecnico, Sede di Milano, Via Saccardo 9, 20134 Milano, Italy
DIN	Deutsches Institut für Normung e.V., Burggrafenstraße 6, D-10772 Berlin, Germany
EN	Comité Européenne de Normalisation - CEN, Rue de Stassart 36, B-1050 Bruxelles, Belgium
GOST or GOST R	Committee of the Russian Federation for Standardization, Metrology and Certification, Leninsky Prospekt 9, Moskva 117049, Russia
IEC	International Electrotechnical Commission, 3 rue de Varembé, P.O. Box 131, CH-1211 Geneva 20, Switzerland
I.S.	NSAI: National Standards Authority of Ireland, Glasnevin, Dublin 9, Ireland
ISO	International Organization for Standardization, Central Secretariat, 1 rue de Varembé, BP 56, CH- 1211 Geneva 20, Switzerland
JIS	Japanese Standards Association (JSA) 1-24 Akasaka 4-Chome Minato-ku, Tokyo 107, Japan

JLPGA	Japanese Liquefied Petroleum Gas Association
NBN	Bureau de Normalisation, Rue de Birmingham, 131, B-1070 Bruxelles Belgium
NBR or NBR NM	Associacao Brasileira de Normas Tecnicas, Avenida 13 de Maio 13/28 Andar CEP 20003-900, Rio de Janeiro RJ, Brazil
NEN	Nederlands Normalisatie-Instituut, Postbus 5059, 2600 GB, Delft, The Netherlands
NF or NFV	AFNOR Assoc. Francaise de Normalisation, 11 Avenue Francis de Pressense, 93571 St Denis la Plaine CEDEX, France
NS	Standards Norway, P.O. Box 242, NO-1326 Lysaker, Norway
ÖNORM	Austrian Standards Institute (ASI), Heinestrasse 38, Postfach 130, A-1021 Wien, Austria
PN	PKN Polish Committee for Standardization, Swietokrzyska 1400-050 Warszawa, Poland
SAC	Standardization Administration of China, No. 9 Madian East Road, Haidian District, Beijing 100088 China
SEV	Verband für Elektro-, Energie- und Informationstechnik, Seefeldstrasse 301, CH-8034 Zürich, Switzerland
SFS	Suomen Standardisoimisliitlo r.y. NT Build 361, Nordtest, Espoo
SIA	Schweizerischer Ingenieur- und Architektenverein, Selnaustrasse 16, CH-8001 Zürich, Switzerland
SN	SNV Swiss Association for Standardization, Buerglistr. 29, CH-8400 Winterthur, Switzerland
SS	Standardiserings-Kommissionen i Sverige, SE-118 80 Stockholm, Sweden
UNE	AENOR: Asociacion Espanola de Normalizacion Certificacion, Fernandez De La Hoz, 52 28010 Madrid, Espana
UNI	Ente Nazionale Italiano di Unificazione, Via Sannio, 2, 20135 Milano, Italy
VDE	Verband der Elektrotechnik Elektronik Informationstechnik e.V., Stresemannallee 15, D-60596 Frankfurt am Main, Germany
VDMA	Verband Deutscher Maschinen- und Anlagenbau e.V., Postfach 71 08 64, D-60498 Frankfurt am Main, Germany

Table 6. Symbols used

Symbol	Significance	Usual unit	SI unit
AH_{air}	Absolute humidity	$g\ kg^{-1}$	1
AH_{air}	Absolute humidity	$g\ m^{-3}$	$kg\ m^{-3}$
a_m	Molecular cross sectional area, occupied by an adsorbed molecule in a complete monolayer	nm^2	m^2
C	BET parameter		1

Table 6. (Continued)

Symbol	Significance	Usual unit	SI unit
AH_{air}	Absolute humidity	$g\ kg^{-1}$	1
AH_{air}	Absolute humidity	$g\ m^{-3}$	$kg\ m^{-3}$
C_{psy}	Psychrometric constant = 0,67 hPa K^{-1} for height up to 500 m and $T_{wet} > 0$ °C	hPa K^{-1}	Pa K^{-1}
m	Sample mass		kg
m_a	Adsorbed mass		kg
M_a	Mol mass of adsorbate		kg mol^{-1}
m_s	Sample mass		kg
m_a/m_s	Specific mass adsorbed		1 = kg kg^{-1}
m_{Air}	Mass of dry air		kg
m_{as}	Asymptotic equilibrium value of mass adsorbed		kg
m_w	Mass of water vapour		kg
$m_{w,max}$	Maximum mass of water vapour that the air could hold at the given temperature		kg
n_a	Specific adsorbed amount		mol kg^{-1}
N_A	Avogadro constant = 6.022 x 10^{23}		mol^{-1}
p	Pressure		Pa
p_0	Saturation pressure	hPa	Pa
p_n	Standard pressure (= 101 325)		Pa
P_{H2O}	Water vapour pressure in air	hPa	Pa
R	Universal gas constant (= 8.314)		J K^{-1} mol^{-1}
RH_{air}	Relative humidity	%	1
SH_{air}	Specific humidity		1
t	time		s
T	Temperature	°C	K
T_n	Standard temperature (= 273.15)		K
T_{dry}	Temperature of dry air	°C	K
T_{wet}	Temperature of the wet surface of the thermometer	°C	K
V_a	Specific adsorbed gas volume		m^3 kg^{-1}
V_{air}	Volume of dry air		m^{-3}
τ	Characteristic time of the adsorption process		s

CONCLUSION

For the determination of atmospheric humidity the methods developed in the 18th and 19th centuries are still predominant. Psychrometry is regarded as a standard method. Increasingly, the measurement is converted to an electrical signal allowing for transmission of results and control of processes. Efforts are directed toward developing water vapour measurements and standards at the part per billion levels and below.

The survey demonstrates that different measuring methods must be applied due to the very different tasks of humidity measurements and due to the great number of materials to be tested. With regard to the disturbances, which are specific to each method, the measurements should be compared to standardised procedures. In many cases, it is necessary to determine water within a mixture of different volatiles. In such cases, water should be identified by chemical methods as a reference. Today Karl Fischer titration is preferred as a reference method. If equilibrium cannot be achieved within reasonable time mathematical extrapolation methods should be applied.

Search for extraterrestrial water is based on remote measurements by analysing the spectrum of electromagnetic radiation either of natural sources or produced by a probe and reflected. Spacious deposits had been detected photographically. In situ investigations are made using conventional methods.

ACKNOWLEDGMENTS

Te paper contains some text, tables and figures published in: Katrin Rübner, Devrim Balköse, Erich Robens: Methods of humidity determination – Part I:. *Journal of Thermal Analysis and Calorimetry* 94 (2008) 3. 669-673, Part II: 94 (2008) 3.675-682. We thank Akademiai Kiado for the permission to reuse the needed parts.

REFERENCES

[1] Robens, E., Rübner, K., Staszczuk, P., Dabrowski, A., *Measurement of water vapour sorption and humidity - A survey on measuring methods and standards.* Annals of the Polish Chemical Society, 2003. 2 / II: p. 670-676.

[2] Robens, E. and K. Rübner, *Gravimetrische Wasserdampfsorptions- und Feuchtemessung an Feststoffen.* GIT Z. Labortechnik, 2003. 47(10): p. 1046-1050.

[3] Robens, E., Rübner, K., Staszczuk, P., Dąbrowki, A., *Measurement of water vapour sorption and humidity. A survey on measuring methods and standards.* ICTAC News, 2005. 38 (1): p. 39-46.

[4] Wernecke, R., *Industrielle Feuchtemessung*. 2003, Weinheim: Wiley - VCH.

[5] Chaplin, M., *Water Structure and Behavior.* 2007, http://WWW.lsbu.ac.uk/water/index.html.

[6] Stillinger, F.H., *Water revisited.* Science, 1980. 209(4455): p. 451-457.

[7] Rübner, K., E. Robens, and D. Balköse, *Methods of humidity determination. Part I: Hygrometry.* J. Therm. Anal. Cal., 2008. 94(3): p. 669-673.

[8] Fraden, J., ed. *Humidity and Moisture Sensors*. Handbook of Modern Sensors. 2004, Springer: New York.

[9] *The Book of Judges.*, in *Bible.* p. Chap. VI, vv. 33-40.

[10] Brunt, D., *Ann. Rep. East Mailing Research Station, Kent, England for 1958*, in 1959. p. 41.

[11] Giles, C.H., *Gideon's Fleece Test. The Earliest Recorded Vapor Phase Adsorption Experiment?* J. Chem. Education, 1962. 39 p. 584.

[12] Gerland, E. and F. Traumüller, Geschichte der physikalischen Experimentierkunst. 1899, Leipzig

[13] Cusanus, N., Idiota de Staticis Experimentis, Dialogus. Codex Cusanus 1456/64 1450, Straßburg Folio 135r.

[14] Alberti, L.B., L'architettura. 1483 / 1485, Padua, Firenze

[15] da Vinci, L., Codex atlanticus - Saggio del Codice atlantico, ed. Aretin. Vol. fol. 249 verso-a + fol. 8 verso-b. 1872, Milano

[16] da Vinci, L., Catalogue "Les Mots dans le Dessin" of the Cabinet des Dessin. 1986, Paris Louvre.

[17] Robens, E. and A. Dąbrowski, Extension of the measuring range of balances. J. Therm, Anal. Cal., 2006. 86(1): p. 17-21.

[18] Gal, S., Die Methodik der Wasserdampf-Sorptionsmessungen. 1967, Berlin: Springer.

[19] Rübner, K., D. Balköse, and E. Robens, Methods of humidity determination. Part II: Determination of material humidity. J. Therm, Anal. Cal., 2008. 94(3): p. 675-682.

[20] Kiefer, S. and E. Robens, Some intriguing items in the history of volumetric and gravimetrc adsorption measurements. J. Therm, Anal. Cal., 2008. 94(3): p. 613-618.

[21] Mecea, V.M., J.O. Carlsson, and R.V. Bucur, Extension of the quartz-crystal-microbalance technique. Sensors and Actuators A, 1996. 53: p. 371-378.

[22] Brokate, T., et al., Survey on mass determination with oscillating systems. Part II: Instruments and weighing of matter from gaseous environment. Journal of Thermal Analysis and Calorimetry, 2003. 71(1): p. 25-29.

[23] Mintova, S., S. Mo, and T. Bein, Chem. Material, 2001. 13: p. 901.

[24] de Saussure, H.B., *Versuch über die Hygrometrie.* Ostwald's Klassiker der exakten Wissenschaften ed. A.J. von Oettingen. Vol. 115 + 119. 1783/1900, Neuchâtel/Leipzig: Engelmann.

[25] Ozturk, S., Balköse, D., Okur, S., Umemura, J, *Effect of humidty on electrical conductivity of zinc stearate nanofilms.* Colloids and Surfaces A: Physicochemical and Engineering Aspects 2007. 302(1-3): p. 67-74.

[26] Balköse, D., Ulutan, S., Özkan, F., Cakicioglu, Celebi, S., Ülkü, S., *Dynamics of water vapor adsorption on humidity-indicating silica gel.* Appl. Surface Sci., 1998. 134: p. 39-46.

[27] Miyazaki, K., M. Hieda, and T. Kano, Ind. Eng. Chem. Res., 1999. 36: p. 88.

[28] Sundaram, R., *Comparative study on morphology and humidity sensitive properties of thick film and disc humidity sensors based on semiconducting* $SnWO_4$-SnO_2 *composites.* Sensors and Actuators. B: Chemical, 2007. 12(4): p. 429-436.

[29] Berkün, D., Balköse, D., Tıhmınlıoğlu, F., Altinkaya, S.A., *Sorption and diffusion of water vapor on edible films.* J. Therm. Anal. Cal., 2008. 94(3): p. 683-686.

[30] Balköse, D., U. Köktürk, and H. Yilmaz, *A study of cobaltous chloride dispersion on the surface of the silica gel.* Appl. Surface Sci., 1999. 147: p. 77-84.

[31] Balköse, D. and S. Ülkü, *Water vapour adsorption and desorption on cobaltous chloride impregnated wool in packed column*, in *Polymers and Composites: Synthesis, Properties, and Applications* R.A. Pethrick, G.E. Zaikov, and D. Horak, Editors. 2007, Nova Science Publishers: New York. p. 35-40.

[32] Dacres, H. and H. Narayanaswamy, Talanta, 2006. 69: p. 631.

[33] AMS, *Glossary of Meteorology.* *http://amsglossary.allenpress.com/glossary/.* 2010, Boston, MA: American Meteorological Society.

[34] Berzelius, J.J., *Chemische Operationen und Geräthschaften*. 3. ed. Lehrbuch der Chemie. Vol. 10. 1841, Dresden, Leipzig: Arnoldsche Buchhandlung.

[35] Tomaschek, R., *Mechanik / Wärmelehre / Akustik*. 12. ed. Grimsehls Lehrbuch der Physik. Vol. 1. 1942, Leipzig, Berlin: Teubner.

[36] Wiehle, M., *Magdeburger Persönlichkeiten*. 1993.

[37] Sprung, A., *: Über die Bestimmung der Luftfeuchtigkeit mit Hilfe des Assmannschen Aspirationspsychrometers,* . Z. Angew. Meteorol., Das Wetter, 1888. 5: p. 105-108.

[38] Hoummady, M., Bonjour, C., Collin, J., Lardet-Vieudrin, F.Martin, G., *Surface acoustic wave (SAW) dew point sensor: application to dew point hygrometry*. Sensors and Actuators B: Chemical, 1995. 27(1-3): p. 315-317.

[39] Park, H., Lee K., Kong G., Seung S., Jung Y., Kim K., *Reliable optical measurement of water vapour in highly scattering environment.* Spectrochimica Acta, 2009. 72(A): p. 510-514.

[40] Kwon, S.Y., J.C. Kim, and B.I. Choi, *Accurate dew point measurement over a wide temperature range using a quatrz crystal microbalance humidity sensor*. Measurement Science and Technology, 2008. 19: p. 1152006-1152013.

[41] Cracknell, A.P. and L. Hayes, *Introduction to Remote Sensing*. 2. ed. 2007, London: Taylor and Francis.

[42] Wikipedia, *Lidar. http://en.wikipedia.org/wiki/LIDAR*. 2007, Wikipedia.

[43] Cimini, D., Westwater, E.R., Gasiewski, A.J., Klein. M., Vaccarini, L., *Temperature and humidity profiling in the Arctic using millimeter-wave radiometry*. IEEE transactions on geoscience and remote sensing, 2010. 48: p. 1381-1387.

[44] Vitruvius Pollio, M., *De Architectura* Vol. II/III. 33 - 14 B.C., Rome.

[45] Mackenzie, R.C., Thermochimica Acta, 1984. 75: p. 251-306.

[46] Eyraud, C. and P. Rochas, *Thermogravimetry and silk conditioning in Lyons. A little known story*. Thermochimica Acta, 1989. 152: p. 1-7.

[47] Hemminger, W.F. and K.-H. Schönborn, *A nineteenth century thermobalance*. Thermochimica Acta, 1980. 39: p. 321-323.

[48] Honda, K., *On a Thermobalance*. Science Reports of the Tôhoku University, Sendai Serie 1, 1915. 4: p. 97-103.

[49] Honda, K., Kinzoku no Kenkyu 1924. 1: p. 543.

[50] Hannay, J.B., J. Chem. Soc., 1877. 32: p. 381.

[51] Warburg, E. and T. Ihmori, *Über das Gewicht und die Ursache der Wasserhaut bei Glas und anderen Körpern.* Annalen der Physik, 1886. 263: p. 481-507.

[52] Emich, F., *Einrichtung und Gebrauch der zu chemischen Zwecken verwendbaren Mikrowaagen.*, in *Handbuch der biochemischen Arbeitsmethoden.*, E. Abderhalden, Editor. 1919: Berlin/Wien. p. 55-147.

[53] Emich, F., *Einrichtung und Gebrauch der zu chemischen Zwecken verwendbaren Mikrowaagen.*, in *Handbuch der biologischen Arbeitsmethoden.*, E. Abderhalden, Editor. 1921: Berlin/Wien. p. 183-269.

[54] Urbain, G. and C. Boulanger, *Sur une balance-laboratoire à compensation électromagnetique à l'étude des systèmes qui dégagent des gaz avec une vitesse sensible.* Compt. rend. , 1912. 154 p. 347-349.

[55] Gast, T., T. Brokate, and E. Robens, *Vacuum weighing*, in *Comprehensive Mass Metrology*, M. Kochsiek and M. Gläser, Editors. 2000, Wiley-VCH: Weinheim. p. 296-399.

[56] Keller, J.U., E. Robens, and C. du Fresne von Hohenesche, *Thermogravimetric and Sorption Measurement Techniques/ Instruments*, in *Characterization of Porous Solids VI.*, F. Rodríguez-Reinoso, et al., Editors. 2002, Elsevier: Amsterdam. p. 387-394.

[57] Sandstede, G. and E. Robens, *Automatisierte Apparatur zur gravimetrischen Bestimmung der spezifischen Oberfläche und der Porengröße.* Chem.-Ing.-Tech., 1962. 34(10): p. 708-713.

[58] Römpp, *Römpp Chemie Lexikon.* CD 1.0 ed. 1995, Stuttgart: Thieme.

[59] *DIN ISO: Internationales Wörterbuch der Metrologie. International Vocabulary of Basic and General Terms in Metrology.* 1984, Berlin: Beuth.

[60] Staszczuk, P., *Studies of the adsorbed water layers on solid surfaces by means of the thermal analysis special technique.* Thermochimica Acta, 1998. 308: p. 147-147.

[61] Liptay, G., *Atlas of Thermoanalytic Curves.* 1971-1977, Budapest, London: Akadémiai Kiadó and Heyden.

[62] *DIN 66138: Isotherme Messung der Aufnahme und Abgabe von Dämpfen durch Feststoffe.* 2008, Berlin: Beuth.

[63] *DIN 50008-1: Klimate und ihre technische Anwendung; Konstantklimate über wäßrigen Lösungen; Gesättigte Salzlösungen, Glycerinlösungen.* 1981, Beuth: Berlin.

[64] Volfkovich, Y.M., et al., *Techniques of standard porosimetry, and possible areas of their use in electrochemistry (Review)*. Soviet Electrochemistry, 1981. 16(11): p. 1325-1353.

[65] Rouquerol, F., J. Rouquerol, and K. Sing, *Adsorption by Powders and Porous Solids*. 1999, San Diego: Academic Press.

[66] Robens, E., Benzler, B., Büchel, G., Reichert, H., Schumacher, K., *Investigation of characterizing methods for the microstructure of cement paste*. Cement and Concrete Research, 2002. 32(1): p. 87-90.

[67] Juhola, A.J., *Manufacture, pore structure and application of activated carbons. Part I*. Kemia-Kemi, 1977. 4(11): p. 543-551.

[68] Aligizaki, K.K., *Pore Structure of Cement-Based Materials*. Modern Concrete Technology 12. 2006, London: Taylor and Francis.

[69] *DIN ISO 9277: Bestimmung der spezifischen Oberfläche von Feststoffen durch Gasadsorption nach dem BET-Verfahren*. 2003, Berlin: Beuth.

[70] Adolphs, J. and M.J. Setzer, *A model to describe adsorption isotherms*. J. Coll. Interface Sci., 1996. 180: p. 70-76.

[71] Adolphs, J. and M.J. Setzer, *Energetic classification of adsorption isotherms*. J. Coll. Interface Sci., 1996. 184: p. 443-448.

[72] Adolphs, J., *Excess surface work - A modelless way of getting surface energies and specific surface areas directly from sorption isotherms*. Appl. Surface Sci., 2007. 253: p. 5645-5649.

[73] Rübner, K., J. Rübner, and G. Knölle, *Porenstruktur von Polyurethan/Kieselsäure-Nanocompositen*, in *XI. Workshop der Porotec GmbH über die Charakterisierung von feinteiligen und porösen Festkörpern, Bad Soden*. 2007, Porotec GmbH, Hofheim.

[74] Rübner, K., Goering, J., Klobes, P., Tschritter, H., *Investigation on the pore structure of silica/polyurethane nanocomposites*, in *Characterisation of Porous Solids VIII*. 2009, *Proceedings of the 8th International Symposium on the Characterisation of Porous Solids, Edinburgh, 10-13 June 2008*, Kaskel, S., Llewellyn, P., Rodriguez-Reinoso, F., Seaton, N. A., Editors, Special Publication No. 318, 2009, The Royal Society of Chemistry, Cambridge, p. 150-158.

[75] *DIN 66134: Bestimmung der Porengrößenverteilung und der spezifischen Oberfläche mesoporöser Feststoffe durch Stickstoffsorption. Verfahren nach Barrett, Joyner und Halenda (BJH)*. 1998, Berlin: Beuth.

[76] Gurwitsch, L.G., *(On the interaction of pysico-chemical forces). In Russian*. J. Russ. Phys. Chem. Soc. Petersburg, 1915. 47: p. 805-827.

[77] Brun, M., et al., *A new method for the simultaneous determination of the size and the shape of pores: The thermoporometry.* Thermochimica Acta, 1977. 21: p. 59-88.

[78] Fischer, K., *Neues Verfahren zur maßanalytischen Bestimmung des Wassergehaltes von Flüssigkeiten und festen Körpern.* Angew. Chemie, 1935. 48(26): p. 394-396.

[79] Schöffski, K., *Die Wasserbestimmung mit Karl-Fischer-Titration.* Chemie in unserer Zeit, 2000. 34(3): p. 170-175.

[80] Scholz, E., *Karl-Fischer-Titration.* 1984, Heidelberg: Springer.

[81] Isengard, H.-D., *Bestimmung von Wasser in Lebensmitteln nach Karl Fischer.* ZFL-Internat. Zeitschr. f. Lebensmittel-Technik, Marketing, Verpackung und Analytik, European Food Science, 1991. 42: p. 1-6.

[82] Isengard, H.-D., *How to determine water in foodstuffs?* Analytix, 2003. 3: p. 11-15.

[83] Schaller, E., *Wassergehalt in Gasen.* LABO, 2007. 38(2): p. 24-26.

[84] *ISO 10110-9: Natural gas. Determination of water.*

[85] *ISO 10110-2: Natural gas: Determination of water by the Karl Fischer method. Part 2: Titrimetric method*, Genève: ISO.

[86] *ISO 10110-3: Natural gas: Determination of water by the Karl Fischer method. Part 3: Coulometric method*, Genève: ISO.

[87] Keidel, F.A., *Determination of water by direct amperometric measurement.* Anal. Chem., 1959. 31(12): p. 2043-2048.

[88] Möhlmann, D., R. Wernecke, and V. Schwanke, *Measurement principle and equipment for measuring humidity contents in the upper martian surface and subsurface*, in *Proc. 37th ESLAB Symposium 'Tools and Technologies for Future Planetary Exploration'.* 2004, ESA: Noordwijk. p. 163-168.

[89] Wang, H.G., Senior P. R., Mann, R., Yang, W.O., *Online measurement and control of solids moisture in fluidized driers.* Chemical Engineering Science, 2009. 64: p. 2893-2902.

[90] Jäntti, O., J. Junttila, and E. Yrjänheikki, *On curtailing the micro-weighing time by an extrapolation method.*, in *Progress in Vacuum Microbalance Techniques*, T. Gast and E. Robens, Editors. 1972, Heyden: London. p. 345-353.

[91] Poulis, J.A., Reichenauer, G., Massen, C.H., Robens, E., *A Jäntti approach for quick calculations of sorption equilibria.* Z. Phys. Chem., 2002. 216: p. 1123-1135.

[92] Robens, E. and C.H. Massen, *On the applicability of Jäntti's method of shortening sorption measurements* Journal of Thermal Analysis and Calorimetry, 2008. 94(3): p. 711-714.

[93] Jäntti, O., J. Junttila, and E. Yrjänheikki, *Mikropunnitusajan Lyhentämisestä Ekstrapolaatiomenetelmällä. (On curtailing the microweighing time by an extrapolation method).* Suomen Kemistilehti, 1970. A 43: p. 214-218.

[94] Möhlmann, D., *Kometen*. Beck'sche Reihe. 1997, München: Beck.

[95] Sokol, A.K., et al., *Geochemistry, petrology and ages of the lunar meteorites Kalahari 008 and 009: New constraints on early lunar evolution.* Geochim. Cosmochim. Acta 2008. 72: p. 4845-4873.

[96] Deiss, B., *Wassereis auf dem Mond http://www.planetenkunde.de /p012/p01205/index.htm*. 2008, AstroLink.de.

[97] Feldman, W.C., et al., *Polar hydrogen deposits on the Moon.* http://lunar.lanl.gov/pubs/2000/Polar_H_Deposits_on_Moon.pdf.

[98] Robens, E., et al., *Investigation of surface properties of lunar regolith - Part IV.* Annales Universitatis Mariae Curie-Skłodowska Lublin, Polonia, Sectio AA Chemia 2008. LXIII: p. 144-168.

[99] Robens, E., Adolphs, J., Bischoff, A., Goworek, J., Kutarov, V.V., Mendyk, E., Schreiber, A., Skrzypiec, K., *Investigation of surface properties of lunar soils.* Zeitschrift für geologische Wissenschaften – Journal of Geological Sciences, 2010.

[100] Saal, A.E., Hauri, E.H., Cascio, M.L., Van Orman, J.A., Rutherford, M.C., Cooper, R.F., *Volatile content of lunar volcanic glasses and the presence of water in the Moon's interior.* Nature, 2008. 454(7201): p. 192-195.

[101] Wikipedia, *Water on Mars. http://en.wikipedia.org/wiki/ Water_on_ Mars*. 2010.

[102] Boynton, W.V., et al., *The Mars Odyssey gamma-ray spectrometer instrument suite.* Space Science Reviews, 2004. 110: p. 37-83.

[103] Boynton, W.V., et al., *Distribution of hydrogen in the near-surface of Mars: Evidence for subsurface ice.* Science, 2002. 297: p. 81-85.

[104] Nägele, E. and H.K. Hilsdorf, *Bestimmung des Wasserzementwertes von Frischbeton (Determination of water cement ratio of fresh concrete)*, ed. D.A.f. Stahlbeton. Vol. 349. 1984, Berlin: Wilhelm Ernst and Sohn.

[105] Rübner, K., Meng, B. and Müller, U. *Investigating the original water content of concrete*, in *Non-destructive evaluation of reinforced concrete structures, Volume 1: Deterioation processes and standard test methods*,

H.-W. Maierhofer, C. Reinhardt, and G. Dobmann, Editors. 2010, Woodhead Publishing Oxford p. 217-241.

[106] Krus, M., *Humidity transport and storage coefficients of porous mineral building materials. Theoretical basiscs and new measuring techniques*. Thesis. 1995, Stuttgart: Universität Stuttgart.

[107] Leschnik, W., *Humidity measurment at building materials – Between classic and modern*, in *Feuchtetag '99*. 1999, DGZfP: Berlin.

[108] Monfore, G.E., *A review of methods for measuring water content of highway components in place*, in *Highway Research record Number 342, Environmental Effects on Concrete*. 1970, Highway Research Board: Washington DC. p. 17-26.

[109] *DIN 66138: Isotherme Messung der Aufnahme und Abgabe von Dämpfen durch Feststoffe*. 2007, Berlin: Beuth.

[110] DIN/VDI, *Umweltmeteorologie. Meteorologische Messungen, Teil 1: Wind, Temperatur, Feuchte und Niederschlag*. DIN-VDI-Taschenbuch 332. 1999, Berlin: Beuth.

[111] *DIN 51006: Thermische Analyse (TA) - Thermogravimetrie (TG) - Grundlagen*. 2005, Berlin: Beuth.

[112] *DIN 51007: Thermische Analyse (TA) - Differenzthermoanalyse (DTA) - Grundlagen*. 1994, Berlin: Beuth.

[113] NIST, *Thermodynamic quantities - Humidity measurements - http://ts.nist.gov/MeasurementServices/Calibrations/Humidity.cfm*. 2007, NIST - National Institute of Standards and Technology.

In: Humidity Sensors
Editor: Christopher T. Okada
ISBN 978-1-61209-246-1

Chapter 2

Environmental and Bio-Medical Applications Using Quartz Crystal Microbalance (QCM) Humidity Sensors Modified with Nano-Assembled Thin Films

Serhiy Korposh, Roman Selyanchyn and Seung-Woo Lee[1]

[1]Graduate School of Environmental Engineering, the University of Kitakyushu, 1-1 Hibikino, Wakamatsu, Kitakyushu 808-0135, Japan

Abstract

Nano-assembled thin films prepared by a layer-by-layer approach on QCM resonators were used as sensitive elements for monitoring relative humidity. Different types of films, polymer-based $(PDDA/PSS)_n$ film and porphyrin-based $(PDDA/TSPP)_n$ and $(PDDA/MnTSPP)_n$ films (where n=5, 10, and 15), showed linear responses to relative humidity changes in the range of 4%–94%. An increase of the number of bilayers deposited on the QCM electrode enhanced the sensitivity to relative humidity. The porphyrin-based films showed ca. 4.5 times higher sensitivity towards relative humidity than the polymer-based films, owing to the presence of the higher number of free sulfonic acid groups. Incorporation of the metal ion into the porphyrin pyrrole ring was not significant in sensitivity. However, all films showed a good reversibility to the stepwise changes of relative humidity and the response and recovery times (t_{90}) were very fast, within 15 sec.

Additional examples of the QCM humidity sensors for environmental and medical sensing are presented. In particular, a respiratory monitoring system based on QCM frequency oscillations was designed and investigated. The developed system, in which the sensor response reflects lung movements, was able to track human breathing parameters such as respiratory rate, tidal volume, and inspiratory/expiratory flow, which were tested on healthy volunteers.

1. INTRODUCTION

Cheap and reliable relative humidity (rH) sensors in the automobile, medical and industrial areas are still sought [1,2]. In particular, the humidity sensing in fuzzy control systems in buildings with high occupant densities for the thermal comfort is highly important and will be useful for the energy and cost efficiency management that becomes a worldwide mission regarding the reduction of carbon dioxide emission [3]. In addition, cheap humidity sensors that are able to operate in the range of 30–80% rH can be synchronized to improve the performance of other gas sensors that are strongly influenced by the surrounding humidity in real environments [4,5].

Many different types of humidity sensors based on impedance [6], capacitance [7], resistance [8], optical [9], surface acoustic wave (SAW) [10] and quartz crystal microbalance (QCM) [11] measurements have been demonstrated. Among them, the QCM device is extremely sensitive and a powerful tool for monitoring mass changes in the nano-gram range [12,13]. The basic principle of QCM sensors is the measurement of the frequency shift as a result of the mass adsorbed on the QCM resonator. How to modify the surface property of the quartz resonator becomes an essential issue for enhancing the range of applications of the QCM sensors [14].

Employment of thin films as sensitive elements is useful to improve the functions of transducers and the layer-by-layer (LbL) method has been used as a powerful surface modification technique [15-17]. The first trial of this method dates back to 1966 when Iler excogitated it [18]. In 1991, Decher and co-workers rediscovered a new film assembly method based on the alternate adsorption of polycations and polyanions onto surfaces [19-21]. This alternate adsorption technique is still expanding its potential because of its versatility and convenience for fabrication of nano-assembled films. In addition, the LbL method has been successfully employed in sensor fabrication with a variety of materials [22-24]. Very recently, vapochromic materials with a chemical formula of [Au2Ag2(C6F5)4L2]n was alternately deposited on an optical fiber in order to create a sensor array for measuring volatile organic compounds [25] and polyelectrolytic multilayers of poly(4-styrenesulfonic acid-co-maleic acid) and poly(allylamine hydrochloride) coupled with a QCM resonator were used for low humidity detection [26,27].

Porphyrins are tetrapyrrolic pigments that are widely present in nature and play an important role in many biological systems [28]. In addition, they can be used as sensitive elements for optical chemical sensors due to their unique spectral features. Moreover, their optical properties can be varied by metallation of the pyrrole ring, which in turn will lead to a higher sensitivity and selectivity to a wider class of chemical compounds [29]. Recently, we have used tetrakis-(4-sulfophenyl)porphine as a sensitive element with a cationic polymer for ammonia sensing [24,30]. The operation of the proposed device was based on the optical change due to the deprotonation of the pyrrole ring by ammonia.

In this chapter, three different types of alternate films deposited onto QCM devices were examined for monitoring relative humidity, in which poly(sodium 4-styrenesulfonate) and two porphyrin derivatives were used as anionic species for the alternate deposition with poly(diallyldimethylammonium chloride) as cationic polymer. The thickness effect of the sensor films on the frequency shift and sensitivity was also investigated. Understanding the effect of relative humidity on the sensor performance is crucial for the reliable sensor system development. From this viewpoint, porphyrin compounds provide the possibility to explain the influence of humidity on the sensor performance based on their extraordinary optical properties.

Additionally, one possible application in medicine of the proposed QCM-based humidity sensor has been demonstrated. In particular, sensor response that reflects lung movements was used to track human breathing parameters such as respiratory rate, tidal volume, and inspiratory/expiratory flow. The assessment of vital signs is a key routine procedure that is performed during all medical examinations, because they reflect a patient's status and are repeated often and even persistent monitoring can be required for bedridden patients. Conventional systems currently used for breath measurement include pneumotachography [31], various kinds of pletysmography [32, 33], air-filled mattresses [34,35] acoustic noise [36], and capnography as technique for blood carbon dioxide content estimation [37, 38]. Minute lung ventilation is partitioned into two representative values; tidal volume (Vt) and respiratory rate (RR) [39]. In healthy subjects, RR is approximately 17 breaths/min and Vt is approximately 400 mL. The establishment of a respiration pattern (together with RR and Vt) in normal individuals is a complicated process; however, recognizing alterations in respiratory patterns is important as an early indicator of disease and in many instances this recognition can lead directly to a diagnosis. Therefore, careful observation of RR and respiration pattern is a crucial part of the physical examination [40]. To the best of our knowledge, direct dynamical tracking of total changes in humid human breath by the use of QCM sensors was not reported previously. In this chapter, we present a novel approach of respiratory monitoring including RR and breathing patterns assessment using the surface modified QCM sensor [41]. The possibility of on-line measuring of the human breathing parameters (RR, Vt, inspiratory/expiratory flow etc.) is demonstrated. As the sensor response is related to a breath chemical composition matrix, the possibility for simultaneous qualitative/quantitative assessment of the breath content is also discussed.

2. Preparation of Humidity Sensing Films

2.1. Surface Treatment of Substrates before Film Deposition

A QCM (9 MHz, USI System) device connected to the multifunction data acquisition system (National Instrument, PXI-6221) and computer for further data treatment was used for frequency shift measurements. Before film deposition, gold-coated QCM electrodes were treated with a piranha solution (96.0% sulfuric acid/30.0–35.5% hydrogen peroxide, 3/1, v/v), rinsed with deionized water, and dried by nitrogen gas. Then, the electrode was immersed in an ethanol solution of 2-mercaptoethanesulfonic acid (10 mM) for 12 hours at 25 °C, followed by rinsing with ethanol and drying with nitrogen gas. Sensor films were assembled onto the 2-mercaptoethanesulfonic acid modified QCM electrode via the following procedure. Prior to film preparation, a 3-cycle precursor film of PDDA/PSS was assembled onto the QCM electrode by alternate immersion into aqueous solutions of poly(diallyldimethylammonium chloride) (PDDA, *M*w: 200,000–350,000, 20 wt% in H_2O, 5 mg/mL) and poly(sodium 4-styrenesulfonate (PSS, *M*w: 70,000, 5 mg/mL) for 20 min, respectively, followed by rinsing in pure water and drying with nitrogen gas after each immersion.

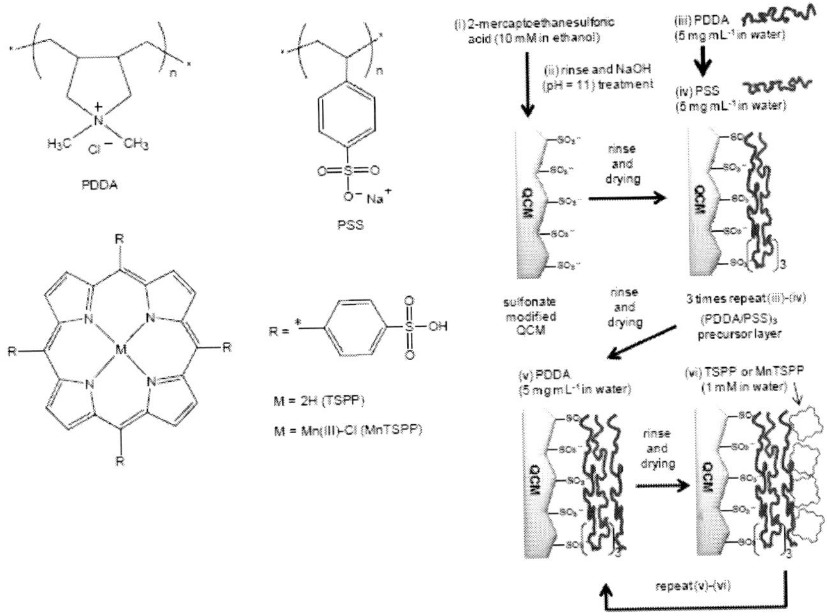

(a) (b)

Figure 1. (a) Chemical structures of polycation (PDDA), polyanion (PSS), and porphyrins (TSPP and MnTSPP) used for film preparation. (b) Schematic illustration of electrostatic layer-by-layer adsorption of PDDA and anionic species onto the QCM electrode.

2.2. Deposition of Sensor Films

The electrostatic LbL deposition was employed for the alternate adsorption of oppositely charged molecular species onto QCM electrodes, as shown in Figure 1. The outermost layer of the precursor film was positively charged with a PDDA layer for the deposition of negatively charged anionic species (AS) such as PSS (5 mg/mL), TSPP (1mM), and MnTSPP (1mM). Three different deposition films of 5-, 10- and 15-cycles (one cycle is a PDDA /AS bilyaer) were prepared by alternate adsorption of PDDA and AS with intermediate rinsing with deionized water and drying by flushing with nitrogen gas. In every case, the alternate film has an outermost surface of AS.

Figure 2 shows QCM frequency shifts due to the alternate adsorption of PDDA and three different anionic species on QCM electrodes. The frequency shifts for all films linearly decreased with the number of deposition cycles, indicating a regular growth of the multilayer films. The average frequency changes of the films with different deposition layers are summarized in Table 1.

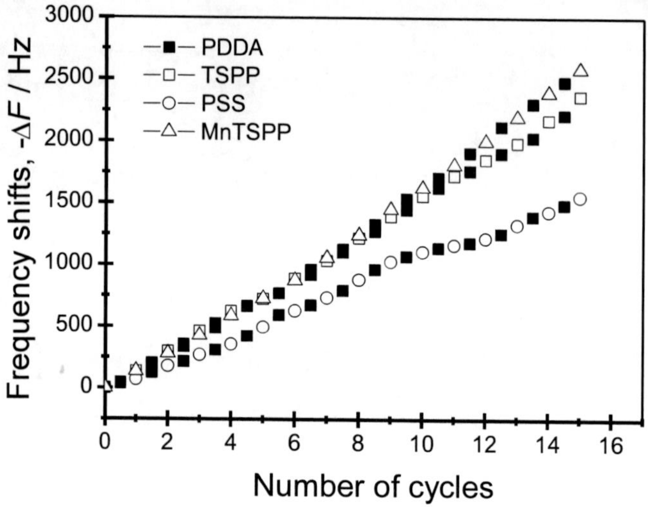

Figure 2. QCM frequency shifts during the alternate deposition of PDDA and anionic species (PSS, TSPP and MnTSPP) on gold-coated QCM electrodes: PDDA/PSS (■/○), PDDA/TSPP (■/□), and PDDA/MnTSPP (■/Δ).

The mass increase due to adsorption can be estimated from the QCM frequency shift by using the Sauerbrey equation [42]. In our system, a frequency decrease of 1 Hz corresponds to a mass increase of ca. 0.9 ng. The thickness (d) of an adsorbed film on one side of a resonator is given by

$$2d\,(\text{Å}) = -\frac{\Delta F\,(\text{Hz})}{1.832\,\rho\,(\text{g/cm}^3)} \tag{1}$$

where ρ is the film density and ΔF is the frequency shift of the QCM. Thus, the thicknesses of each layer can be estimated as follows: in the current setup, a QCM frequency decrease of 1 Hz corresponds to a thickness increase of 0.233/ρ Å, where ρ is the density of the adsorbed film. When the density of the polyelectrolytes (PDDA and PSS) used in this work is assumed to be 1.2 g/cm^3

[43], the ΔF values of 51±21 and 53±17 Hz correspond to a thickness increase of 10±2 and 10±4 Å for one layer of the 15-cycle PDDA/PSS film.

Table 1. Frequency shift and thickness of the individual layer of 15-cycle sensor films

films	frequency shift (Hz)		thickness (Å)		
	PDDA	AS a	PDDA	AS	total
(PDDA/PSS)15	51 ± 21	53 ± 17	10 ± 2	10 ± 4	300
(PDDA/TSPP)15	48 ± 11	109 ± 22	9 ± 2	20 ± 4	435
(PDDA/MnTSPP)15	79 ± 22	92 ± 15	15 ± 2	16 ± 4	465

respectively,orphyrin sensor films
99999999999999999999999999999999999999999999999999999999999
99999999999999999999999999999999999999999999999999999999999
99999999999999999999999999999999999999999999999999999999999
99 [a] AS means anionic species (PSS, TSPP, and MnTSPP).

Interestingly, the average thickness value of the PDDA layer varies depending on the anionic species used for film preparation. The adsorbed mass of PDDA was estimated to be 46±19, 43±10 and 71±20 ng for PSS, TSPP and MnTSPP anionic species, respectively. This difference could be owing to the strength of the electrostatic interaction between PDDA and AS components. Thus, the thickness of the PDDA layer decreases in the order of MnTSPP > PSS ≈ TSPP. The variance of the adsorbed mass of PDDA between TSPP and MnTSPP may be explained by the formation of different types of aggregates when they were assembled on the PDDA layer.

In order to prove this hypothesis, UV-vis spectra of the $(PDDA/TSPP)_{15}$ and $(PDDA/MnTSPP)_{15}$ films were measured, Figure 3. Similarly to the QCM results, the absorbance linearly increased with the increment of the number of layers deposited onto the quartz substrate. The difference in the absorbance value between TSPP and MnTSPP is due to their different molar extinction coefficients [44]. In general, TSPP has two different forms in water and one of them is due to protonation of the porphyrin free base. This can be confirmed from UV-vis spectra of TSPP, in which two different absorption peaks of TSPP has are observed in the Soret band region: non-protonated (peak at 415 nm) and protonated (peak at 434 nm and an additional Q-band at 645 nm) pyrrole rings. The presence of two Soret bands (407 and 492 nm) and a Q-band (702 nm) reveals that the TSPP layer in the film includes a mixed structure of monomer and *J*-aggregate that is characterized as an edge-to-edge form of the constituent TSPP on PDDA layer [24,30,44]. Consequently, some of the sulfonic groups are involved in the formation of the *J*-aggregation and the remaining ones are available for the interaction with PDDA. In contrast, MnTSPP is adsorbed in the form of monomer [44] and this MnTSPP outermost layer, therefore, will offer stronger negative surface charge than TSPP, leading to the higher adsorbed mass of PDDA.

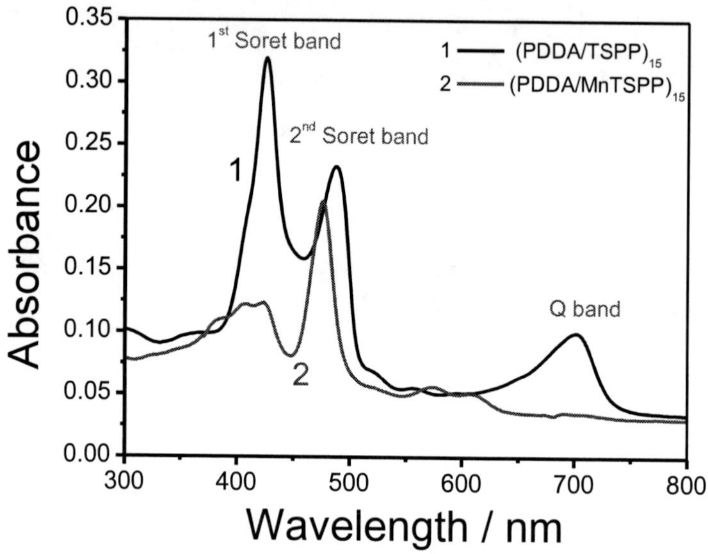

Figure 3. UV-vis spectra of $(PDDA/TSPP)_{15}$ and $(PDDA/MnTSPP)_{15}$ films.

Similarly, the thickness of the negatively charged species of TSPP and MnTSPP was calculated to be 20±4 and 16±4 Å, respectively, using the TSPP and MnTSPP bulk densities (1.3 g/cm^3) [45]. The porphyrin-based films are thicker as compared with the polymeric film of PDDA/PSS that is formed only by the surface charge difference between species used for the film preparation. The average thickness of the PSS layer (10±2 Å) was almost same as that of the PDDA layer, corresponding well with the fact that both components are linear polyelectrolytes. The average thicknesses of TSPP and MnTSPP are very close to their molecular dimension (18×13 Å) [45], suggesting that they were assembled as monolayer onto the PDDA layer, although the TSPP layer is slightly thicker than that of MnTSPP. This small difference between TSPP and MnTSPP may be attributed to the *J*-aggregation of TSPP in the film.

3. Relative Humidity Response

Humidity measurements were undertaken using a set-up shown in Figure 4. The QCM electrodes coated with the different sensor films were placed inside a measurement chamber of ca. 50 mL capable to accommodate four QCM resonators simultaneously. Different relative humidities were set up inside the chamber by admitting humidified air at a flow rate of 1 L/min, controlled by a flow controller (FC). Synthetic dry air was humidified by flowing through a humidifier, as shown in Figure 4. The reversal of the response was studied by flushing the measurement chamber with dry air. A humidity logger (Hygrochron, KN Laboratories: relative humidity range of 0–95%; accuracy ±5% at 25 oC in the range of relative humidity 20–80% and reading resolution 0.1%) was used to measure humidity inside the measurement chamber.

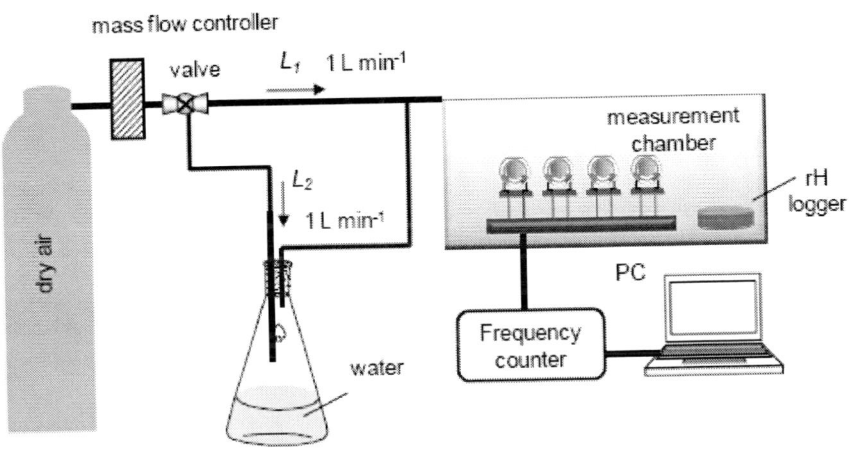

Figure 4. Measurement set-up used for relative humidity sensing.

In order to investigate the basic sensitivity of the sensor films to relative humidity, three different 10-cycle films, $(PDDA/PSS)_{10}$, $(PDDA/TSPP)_{10}$ and $(PDDA/MnTSPP)_{10}$, were tested in the relative humidity range of ca. 4% to 55%. Typical frequency shifts due to the increase and decrease of relative humidity are shown in Figure 5. Stepwise increase of relative humidity from ca. 4% to 55% resulted in the corresponding frequency decrease of 11, 55, 225 and 245 Hz for the non-modified QCM electrode and modified QCM electrodes with $(PDDA/PSS)_{10}$, $(PDDA/TSPP)_{10}$ and $(PDDA/MnTSPP)_{10}$, respectively. The response of the non-modified electrode was very small in time in respect to the relative humidity change measured using a commercial humidity meter. The modified electrodes, especially with TSPP and MnTSPP, showed bigger responses in frequency compared with that of the non-modified bare electrode, indicating that TSPP and MnTSPP used as sensitive elements can be utilized for efficient binding of water molecules. In addition, the response and recovery times are much faster compared with the commercial rH sensor that requires a longer time (150 sec) to return to the base line, while the recovery time of the QCM devices is within 15 seconds.

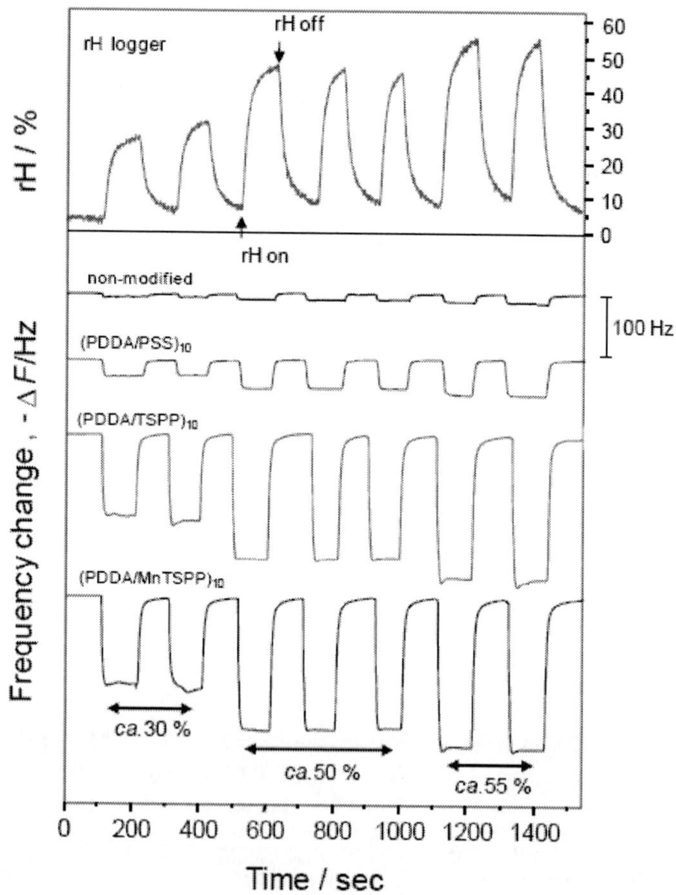

Figure 5. Responses of non-modified and 10-cycle film deposited QCM electrodes to stepwise relative humidity changes.

4. Thickness Dependence

Film thickness is a significant parameter that helps to understand the basic properties of the sensor films at relative humidity change. Table 2 compares the QCM response of the PDDA/PSS, PDDA/TSPP and PDDA/MnTSPP films with different deposition cycles to relative humidity. In all cases, the increase of the films thickness from 5 to 15 cycles resulted in the bigger frequency shift, suggesting the more efficient binding of water molecules due to higher number of available adsorption sites. The 5, 10, 15-cycle PDDA/PSS films showed frequency changes of 44, 67 and 170 Hz for the relative humidity change from ca. 4% to 90%, respectively. For the PDDA/TSPP thin films, the same increment of deposition cycles from 5 to 15 resulted on the 3 fold increase in sensitivity. This thickness dependency of the QCM response suggests that not only the outermost layer but the entire volume of film can be involved in the water sorption.

Table 2. Sensor parameters to relative humidity for different sensor films

films	frequency shift a (Hz)	sensitivity, Hz/relative humidity (%)	linear range (%)
(PDDA /PSS)5	44 ± 1	0.45 ± 0.02	4-90
(PDDA /PSS)10	67 ± 1	0.70 ± 0.02	4-90
(PDDA /PSS)15	170 ± 1	1.69 ± 0.07	4-90
(PDDA /TSPP)5	201 ± 1	2.17 ± 0.04	4-90
(PDDA /TSPP)10	461 ± 1	4.97 ± 0.13	4-90
(PDDA /TSPP)15	639 ± 1	6.84 ± 0.15	4-90
(PDDA /MnTSPP)5	180 ± 1	1.99 ± 0.07	4-90
(PDDA /MnTSPP)10	525 ± 1	5.43 ± 0.17	4-90
(PDDA /MnTSPP)15	982 ± 1	6.39 ± 0.20	4-70

[a] Frequency shift for the relative humidity change from 4% to 90%.

The porphyrin-based films have ca. 4.5 times higher sensitivity than the PDDA/PSS films, Table 2. Interestingly, the 5 and 10-cycle PDDA/TSPP and PDDA/MnTSPP films showed very similar behaviors in sensitivity towards relative humidity and they have linear responses in the wide relative humidity range (4-90%). On the other hand, the 15-cycle PDDA/MnTSPP film showed a different trend that has a linear response in the range from 4 to 70% and a non-linear behavior in the range from ca. 70 to 90%, while all PDDA/TSPP films show linear responses in the whole humidity range [41]. This non-linear response of the 15-cycle PDDA/MnTSPP film could be attributed to the condensation of water inside the film at higher relative humidity.

5. HUMIDITY SENSING MECHANISM

The frequency decrease is owing to the mass increase of adsorbed water molecules on the surface of the QCM resonator, according to the Sauerbrey equation (1). Thus, the consecutive increase of the mass load on the non-modified and PSS modified (10-cycle PDDA/PSS film) QCM electrodes is estimated to be 8.0 and 49.5 ng, respectively, for the relative humidity change from 4 to 55%. As described above, the $(PDDA/TSPP)_{10}$ and $(PDDA/MnTSPP)_{10}$ films show a 4–4.5 fold increase in the mass of the adsorbed water as compared with that of the $(PDDA/PSS)_{10}$ film for the same relative humidity change. The sensitivity of the sensor films to relative humidity seems to be closely related to what kinds of AS were assembled in the films.

The adsorption of water molecules is negligible in the non-modified QCM electrode and relatively small in the PSS-involved QCM electrodes. In contrast, the water adsorption of the porphyrin-based sensor films was remarkably enhanced and it seems more likely to be because of the interaction between water and free sulfonic acid groups of TSPP or MnTSPP [46]. PSS that was assembled in the PDDA/PSS films is composed of sodium salt of the sulfonic acid and therefore can not participate in the water adsorption. It is known that TSPP, in general, has equilibrium with water vapor of the surrounding atmosphere at room temperature [46,47]. That is, the increase of the surrounding relative humidity leads to the higher water content present in TSPP. As can be seen from the absorbance spectra in Figure 3, TSPP in the film predominantly exists in the monomeric form. Thus, water may be bound via hydrogen bonding to the free sulfonic acid groups of TSPP [46] and the increase of humidity will cause the deprotonation of the TPPS sulfonic groups. The dissociated protons can be transferred to the pyrrole ring, leading to the increase of absorbance in the Soret and Q-band regions (data not shown). Schematically, this can be illustrated as shown in Scheme 1. In addition, the increased water adsorption of the porphyrin-based sensor films may be also related to the increased porosity of the films caused by the presence of nano-sized pillar structures permitting better access of analyte water. This adsorption behavior appears more remarkable in the MnTSPP film because of its non-linear performance for humidity.

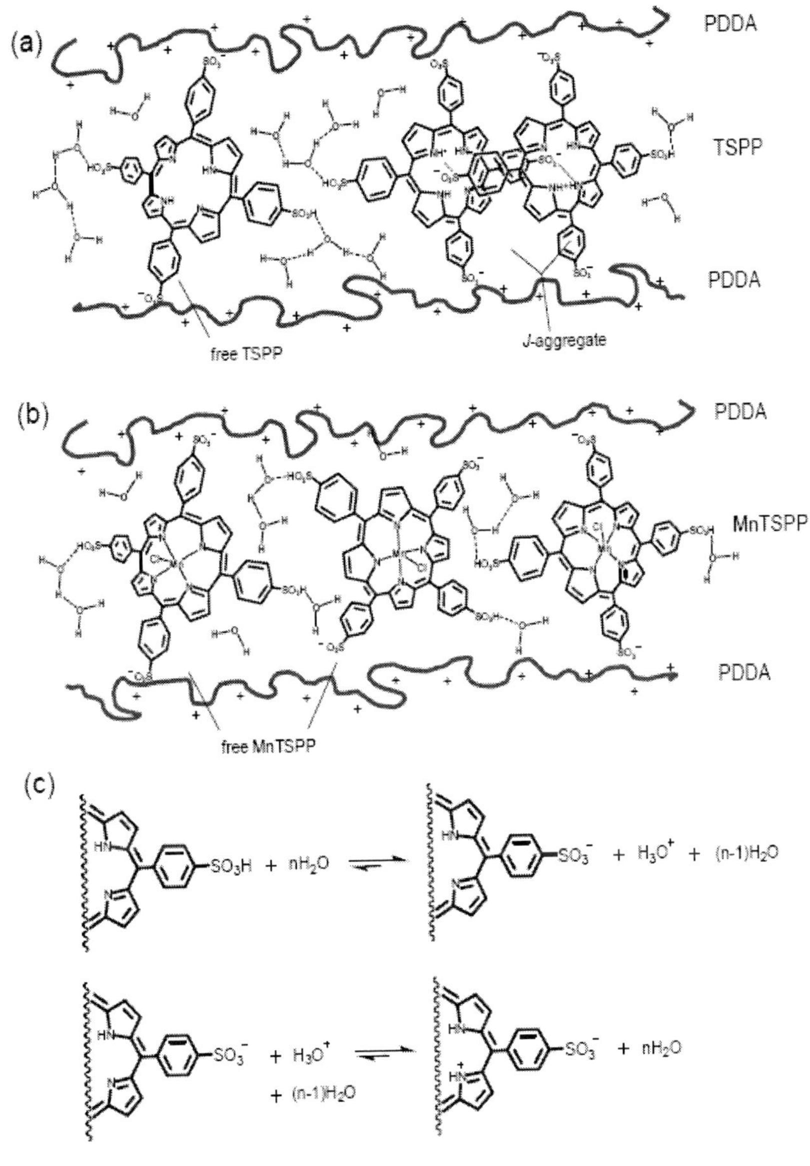

Scheme 1. Schematic illustration of the interaction between water and porphyrin based films, (a) PDDA/TSPP and (b) PDDA/MnTSPP, and (c) chemical reaction between a free sulfonic group and water molecules.

On the other hand, the non-linear sensitivity of the 15-cycle MnTSPP film at higher relative humidity will be attributed to the presence of spaces between the adsorbed MnTSPP molecules, since MnTSPP is assembled as monomeric form on the PDDA layer, while a part of TSPP creates a *J*-aggregated structure that limits space for the water penetration into the film. Anyway, both the TSPP and MnTSPP films have similar amounts of free sulfonic acid groups, although the average frequency of TSPP during the film preparation is slightly larger than that of MnTSPP (see Table 1).

6. Real-Life Measurements and Film Stability

In order to study the performance of the QCM humidity sensors in the real-life application, four QCM electrodes of different types were chosen and placed in the room for one day. The results of this study are shown in Figures 6a and 6b, which were measured in May and February, respectively, with a 10 month interval between both measurements.

It is clear that the response of all QCM sensors used follows the relative humidity values measured using the conventional humidity meter (line 5 in Figures 6a and 6b). In particular, the sensitivity of the porphyrin-based QCM sensors is much higher than that of the non-modified and PSS modified ones. This corresponds well with the results shown in Table 2 for the given relative humidity range. It should be noted that the response of the MnTSPP film is not linear in all range of relative humidity [41]. In our case, however, the range of humidity in the room was from 36% to 65%, in which the film shows a perfectly linear response [41].

An interesting trend was observed for the first trial (measured in May): the frequency started to decrease when the air conditioner in the room was switched off, leading to the increase of the humidity (line 5, Figure 6a). A similar effect, with an opposite sign, was observed when the measurement was conducted in January (Figure 6b): when the air conditioner was switched on, the relative humidity of the commercial humidity logger and the QCM frequency simultaneously started to decrease. Interestingly, the temperature change during the measurement was only $\pm 2^{o}$C and this did not affect on the sensor response (see line 6, Figure 6b). One can also notice that the QCM response is faster and shows better resolution as compared with the commercial sensor. For example, the QCM sensors show more sensitive responses to the relative humidity change, while the commercial humidity sensor gives an obtuse humidity change of ca. ±2% at around 600 min (in circled area). This indicates that much definite humidity information can be obtained from the QCM response.

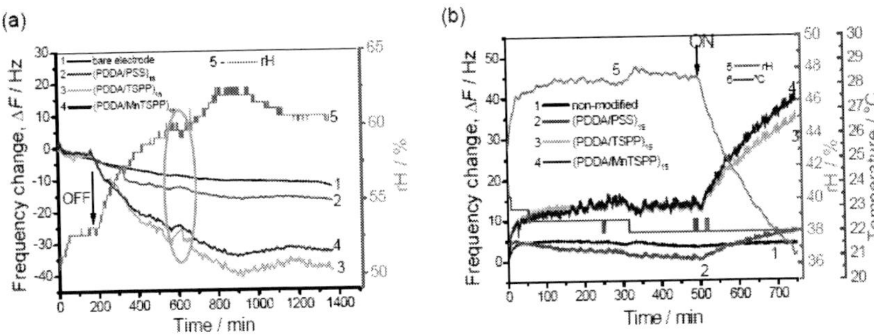

Figure 6. Real humidity responses of non-modified and 15-cycle film deposited QCM electrodes: (a) measured in 02 May and (b) 20 January; the arrows indicate switch on/off the air-conditioner.

Another interesting observation is that at slow relative humidity change the response of the TSPP and MnTSPP films are almost the same (Figure 6a), however, at fast humidity change the TSPP film showed faster and more sensitive response. Perhaps, TSPP is more sensitive to additional environmental changes due to temperature and some chemicals because of its higher reactivity. However, there is no clear evidence for such explanation in this state and additional studies are needed with more planned conditions. On the other hand, the QCM responses were reproducible even after 10 months when the same electrodes were used for demonstrating long time stability. These results suggest that QCM sensors with nano-assembled layers have a high potential for applications in buildings for measurements of the relative humidity change in real time. Additionally, they can be used for monitoring many precise relative humidity changes, owing to their fast response and high resolution.

The thermal stability of porphyrin was measured after annealing the $(PDDA/TSPP)_{10}$ film at 150 °C for 1 hour, showing that the TSPP film is functioning even after annealing at a high temperature (data not shown). It is known that the annealing of TSPP up to 300°C leads exclusively to the loss of the bound water [47]. Interestingly, the film sensitivity could be slightly improved as compared with the response measured before annealing, which can be owing to the purification of the film at higher temperature and availability of the higher amount of binding sites for water. Consequently, this procedure will be useful to refresh the sensor films used in this study.

7. Respiratory Measurements Based on Humidity Sensing

For the breath measurements, two QCM sensors coated with the different porphyrin-based films and one additional uncoated electrode were used. A schematic illustration of the experimental set-up used for the breathing analysis is shown in Figure 7. The measurement chamber was connected to a QCM frequency control device (NAPICOS PSA10A, Nihon Dempa Kogyo Co., Ltd., Japan) with a sampling rate equal to 1 second. A face mask with an inhalation filter (model 3100, 3M, Japan) modified with an additional outlet for exhaled breath transport to the measurement chamber containing the QCM electrodes (Figure 7b) was employed to monitor the breathing oscillation patterns. As additional key components, the following were employed: a mini-pump with regulated flow in range of 0.05–1.5 L min^{-1} (MP-Sigma 30, Sibata, Japan) and humidity logger (DS1923, KN Laboratories, Japan) located in the proximity to QCM electrodes as shown in Figure 7a. Consequently, the humidity of gaseous mix delivered by pumping can be measured with high precision.

In proposed measuring setup, exhalation is done not directly into the chamber, but breath is delivered using a pump that provides continuous, uniform and stable breath flow into the measurement chamber. Additionally, it allowed reducing the breath humidity by dilution of the exhaled breath with the ambient air to avoid the direct condensation of water on the electrodes. Another important feature is that during the inhale, the chamber is ventilated by ambient air enabling signal recovery.

The following protocol was used in order to check possibilities of the breathing monitoring using the QCM system. Participants were asked to breathe normally in a sitting position while wearing a face mask. Breathing through the nose (usual breathing) was required for 5–10 min while the sensor response was recorded on a computer. For a check of reproducibility, participants were asked to repeat the breathing procedure 3 times with rest intervals (ca. 5 min) for sensor recovery to the baseline.

(a)

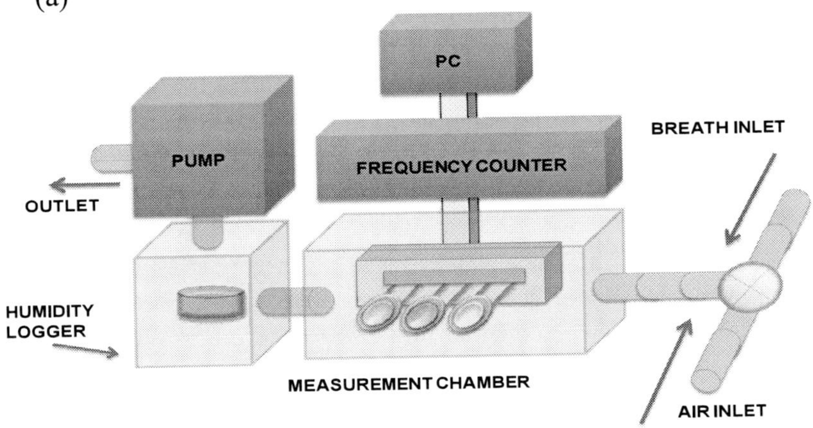

Figure 7. Continued on next page.

(b)

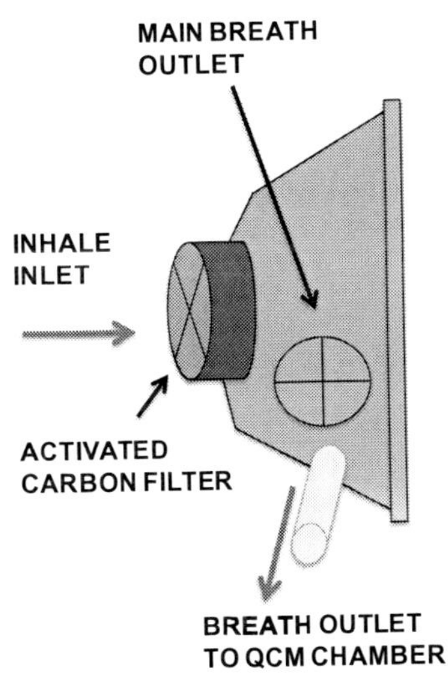

Figure 7. (a) Experimental set-up used for breath monitoring by QCM sensors and (b) schematic illustration of the face mask used for breathing during the experiment.

Figure 8a shows typical QCM sensor response to 3 consecutive human breaths, which consists of two main features: (i) absolute frequency decrease (ΔF) due to the breath delivery into the chamber, which is induced by the increase of humidity and consequent adsorption of water on the QCM electrodes; (ii) amplitude of breath oscillations (A) that are attributed to the inhales and exhales and are observed after frequency saturation due to the delivered breath, Figure 8b. Generally, the sensor showed good reproducibility for three consecutive breath experiments using the blank and porphyrin-modified electrodes (Figure 8a). However, a small decrease in the absolute frequency change (ca. 25 Hz) was observed for the TSPP modified QCM electrode between the first, second and third measurements. The difference in relative humidity values at each measurement was assumed to be a possible reason for such a response. For all three consecutive tests, the relative humidity has almost the same value (mean value 84.3±0.9%), so that its influence cannot explain the decrease in the absolute frequency change. On the other hand, a perfectly reproducible result was observed in the case of the $(PDDA^+/MnTSPP^-)_{15}$ electrode, and thus it is considered that the absolute frequency change was mainly induced by the influence of humidity [41]. This sensor behavior indicates complex processes that are attributed to the interactions between the breath and the sensitive coatings and could not be explained by a baseline shift (when some molecules are attached and not properly removed during air blowout), because the baseline is stable. However, the results presented show good reproducibility for the signals from the same person. The small change in the frequency shift for reproducible channels could be in range of the relative measurement error, and therefore, this has no influence on the breathing oscillations and assessment of the breathing rate.

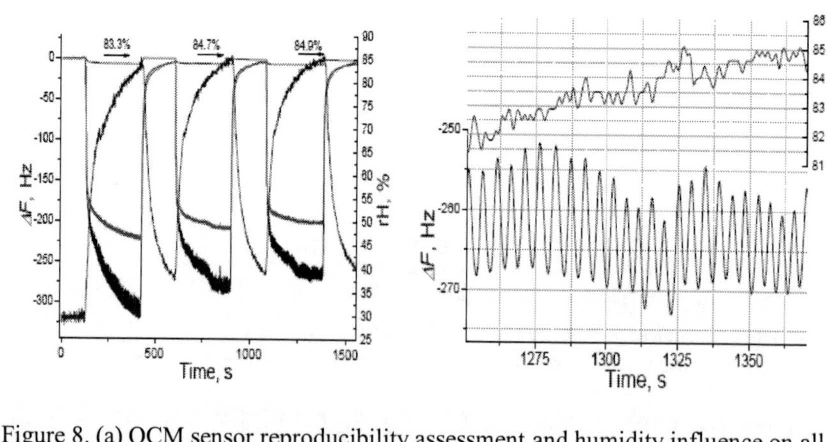

Figure 8. (a) QCM sensor reproducibility assessment and humidity influence on all channels response to breathing: green line, blank electrode; red line, (PDDA+/MnTSPP-)15; black line, (PDDA+/TSPP-)15; blue line, humidity change during breathing (right y-axis). (b) Enlarged view of the sensor response compared to commercial humidity logger data.

Table 3. Assumed relationship between QCM sensor responses and respiratory parameters*

QCM sensor response	Respiratory parameters [unit]
Absolute frequency shift, ΔF, Hz	Breath chemical composition including humidity
Frequency of signal oscillations, RR, min^{-1}	Respiratory rate [min^{-1}]
Amplitude of signal oscillations, V_t, Hz**	Tidal volume [L]
Amplitude of oscillation at deep inhale, V_{IC}, Hz**	Inspiratory lung capacity [L]
Maxima on derivative signal at normal breathing, f_I, Hz s^{-1}**	Inspiratory flow [L s^{-1}]
Minima on derivative signal at normal breathing, f_E, Hz s^{-1}**	Expiratory flow [L s^{-1}]
Maxima on derivative signal at forced inhale, f_{Imax}, Hz s^{-1}**	Peak inspiratory flow [L s^{-1}]
Minima on derivative signal at forced exhale, f_{Emax}, Hz s^{-1}**	Peak expiratory flow [L s^{-1}]
Positive derivative time period, t_I, s	Inspiration time [s]
Negative derivative time period, t_E, s	Expiration time [s]

* Assumption is based on the comparison of the measured QCM signal with typical breathing pattern [29], comparison with the reference methods is further stage of the research

** Probably also related to breath chemical composition.

Additional comparison of the sensor response and humidity logger data is showed in Figure 8b (enlarged view). This data shows that it is not possible to track human breath movements by the recording humidity data by the commercial device used in this work. However, QCM with modified surface gives us the opportunity to track breathing movements in a convenient way with the real time registration; this could be owing to faster response time of the QCM–based humidity sensor as compared to commercial device, as was demonstrated in *section 6*.

Interpretation of the signals obtained from the system is important and response features that might be related to the clinical diagnostic parameters that are listed in Table 3. The oscillating signals of normal breath include V_t and V_{IC} breathing parameters, where V_{IC} indicates a response to the deep inhalation/forced exhalation. The speed of the signal changes is represented in a derivative plot, including positive and negative maximum peaks (f_{Imax} and f_{Emax}) that respect to the peak inspiratory and expiratory flows. However, all sensor parameters should be calibrated with conventional breath measuring systems. One parameter that can be readily assessed is *RR*, which can be easily counted from the differential graph.

CONCLUSION

In this chapter, nano-assembled thin films deposited on QCM resonators were used as sensitive elements for monitoring relative humidity. The sensing mechanism is based on the frequency change induced by mass load owing to chemical interaction of water molecules to the sensor films. Porphyrin-based films of TSPP and MnTSPP, because of their unique chemical structures, showed higher sensitivity as compared with PSS polymer-based films in the relative humidity range of 4%–90%. An increase of the number of bilayers deposited on the QCM resonator led to the ca. 3 times increase in sensitivity towards relative humidity for all films. Good reversibility and fast response and recovery times (<15 sec) to stepwise changes of relative humidity were obtained, indicating the possibility to apply the current system for practical applications ranging from usual humidity measurements to the check of air conditioning systems in buildings. This assumption was proved by monitoring relative humidity in the room for one day with a 10 month interval between measurements.

Additionally, human breath signals were investigated using a QCM three-channel sensor array employing electrodes modified with porphyrin-based functional coatings. The assessment of the useful vital sign parameter, RR, was achieved with a simple experimental setup involving two coated and one blank electrodes, while in principle, only one modified QCM electrode was sufficient for RR measurement. Additional useful responses, Vt of the lungs, and expiratory and inspiratory flows, can be simultaneously obtained after calibration with conventional techniques. We believe that the current approach and system will be useful in various places where precise humidity sensing is needed.

ACKNOWLEDGMENTS

This work was supported by MEXT via the 2nd Kitakyushu Knowledge-based Cluster Project. We would like to thank Mr. Shunichi Wakamatsu (Nihon Dempa Kogyo Co. Ltd.) for his helpful discussion.

REFERENCES

[1] N. Yamazoe, Y. Shimizu, Humidity sensors: principles and applications, *Sens. Actuators B*. 10 (1986) 379–398.

[2] F. Pascal-Delannoy, B. Sorli, A. Boyer, Quartz Crystal Microbalance QCM used as humidity sensor, Sens. Actuators B. 84 (2000) 285–291.

[3] A.B. Shepherd, W.J. Batty, Fuzzy control strategies to provide cost and energy efficient high quality indoor environments in buildings with high occupant densities, Building. *Serv. Eng. Res. Tecnol.* 24 (2003) 35–45.

[4] J. Courbat, D. Briand, J. Damon-Lacoste, J. Wollenstein, N.F. de Rooij, Evaluation of pH indicator-based colorimetric films for ammonia detection using optical waveguides, *Sens. Actuators B*. 143 (2009) 62–70.

[5] S.-I. Ohira, P.K. Dasgupta, K.A. Schug, Fiber Optic Sensor for Simultaneous Determination of Atmospheric Nitrogen Dioxide, Ozone, and Relative Humidity, *Anal. Chem*. 81 (2009) 4183–4191.

[6] P.M. Faia, C.S. Furtado, A.J. Ferreira, AC impedance spectroscopy: a new equivalent circuit for titania thick film humidity sensors, *Sens. Actuators B*. 107 (2005) 353–359.

[7] R.K. Ralston, C.F. Klein, P.E. Thoma, D.D. Denton, A model for the relative environmental stability of a series of polyimide capacitance humidity sensors, *Sens. Actuators B*. 34 (1996) 343-348.

[8] P.M. Faia, C.S. Furtado, A.J. Ferreira, Humidity sensing properties of a thick-film titania prepared by a slow spinning process, *Sens. Actuators* B. 101 (2004) 183–190.

[9] B.D. Gupta, Ratnanjali, A novel probe for a fiber optic humidity sensor, *Sens. Actuators B*. 80 (2001) 132-135.

[10] M. Penza, G. Cassano, Relative humidity sensing by PVA-coated dual resonator SAW oscillator, *Sens. Actuators* B. 68 (2000) 300–306.

[11] Y. Zhang, K. Yu, R. Xu, D. Jiang, L. Luo, Z. Zhu, Quartz crystal microbalance coated with carbon nanotube films used as humidity sensor, *Sens. Actuators A*. 120 (2005) 142–146.

[12] X. Wang, J. Zhang, Z. Zhu, J. Zhu, Humidity sensing properties of Pd2+-doped ZnO nanotetrapods, *Applied Surface Science*. 253 (2007) 3168–3173.

[13] C. Lu, A.W. Czanderna, Applications of Piezoelectric Quartz Crystal Microbalances, Elsevier, Amsterdam, 1984.

[14] R. Paolesse, C. Di Natale, A. Macagnano, F. Davide, T. Boschi, A. D'Amico, Self-assembled monolayers of mercaptoporphyrins as sensing material for quartz crystal microbalance chemical sensors, *Sens. Actuators B*. 47 1998 70–76.

[15] Ichinose, K. Fujiyoshi, S. Mizuki, Y. Lvov, T. Kunitake, Layer-by-layer assembly of aqueous bilayer membranes on charged surfaces, *Chem. Lett*. 25 (1996) 257–258.

[16] S.-W. Lee, I. Ichinose, T. Kunitake, Molecular imprinting of azobenzene carboxylic acid on a TiO_2 ultrathin film by the surface sol-gel process, *Langmuir*. 14 1998 857–2863.

[17] K. Ariga, J.P. Hill, M.V. Lee, A. Vinu, R. Charvet, S. Acharya, Challenges and breakthroughs in recent research in self-assembly, *Sci. Technol. Adv. Mater*. 9 (2008) 1–96.

[18] R.K. Iler, Multilayers of colloidal particles, *J. Colloid Interface Sci*. 21 (1966) 569–594.

[19] G. Decher, J.D. Hong, Buildup of ultrathin films by a self-assembly process: Consecutive adsorption of anionic and cation bipolar amphiphiles and polyelectrolytes on charged surface, *J. Ber Bunsenges Phys. Chem*. 95 (1991) 1430–1434.

[20] Y. Lvov, G. Decher, H. Mchwald, Assembly, structural characterization and thermal behavior of layer-by-layer deposited ultrathin films of polyvinylsulfonate and polyallylamine, *Langmuir*. 9 (1993) 481–486.

[21] J.J. Ramsden, Yu.A. Lvov, G. Decher, Optical and X-ray structural monitoring of molecular films assembled via alternate polyion adsorption, *Thin solid Films*. 254 (1995) 246–251; 261 (1995) 343–344.

[22] J.M. Corres, F.J. Arregui, I.R. Matías, Sensitivity optimization of tapered optical fiber humidity sensors by means of tuning the thickness of nanostructured sensitive coatings, *Sens. Actuators B*. 122 (2007) 442–449.

[23] J. Spadavecchia, G. Ciccarella, P. Siciliano, S. Capone, R. Rella, Spin-coated thin films of metal porphyrin–phthalocyanine blend for an

optochemical sensor of alcohol vapours, *Sens. Actuators B.* 100 (2004) 88–93.

[24] S. Korposh, S. Kodaira, W.J. Batty, S.W. James, S.-W. Lee, Nano-assembled thin film gas sensor. II. An intrinsic high sensitive fiber optic sensor for ammonia detection, *Sensor and Materials.* 21(4) (2009) 179–189.

[25] C. Elosua, C. Bariain, I.R. Matias, F.J. Arregui, E. Vergara, M. Laguna, Optical fiber sensing devices based on organic vapor indicators towards sensor array implementation, *Sens. Actuators* B. 137 (2009) 139–146.

[26] P.-G. Su, K.-H. Cheng, Self-assembly of polyelectrolytic multilayer thin films of polyelectrolytes on quartz crystal microbalance for detecting low humidity, *Sens. Actuators* B. 142 (2009) 123–129.

[27] P.-G. Su, K.-H. Cheng, Layer-by-layer assembly of mica and polyelectrolyte for use in low-humidity sensor, *Sens. Actuators* B. 137 (2009) 555–560.

[28] K.M. Kadish, K.M. Smith, R. Guilard, The Porphyrin Handbook, Academic Press, San Diego, 2003.

[29] A. Morales-Bahnik, R. Czolk, H.J. Ache, An optochemical ammonia sensor based on immobilized metalloporphyrins, *Sens. Actuators* B. 18–19 (1994) 493–496.

[30] S. O. Korposh, N. Takahara, J.J. Ramsden, S.-W. Lee, T. Kunitake, Nano–assembled thin film gas sensors. I. Ammonia detection by a porphyrin–based multilayer film, *Journal of Biological Physics and Chemistry.* 6 (2006) 125–133.

[31] J. H. Lecky, J. K. Denlinger, A. J. Ominsky, A new disposable spirometer, *Anesthesiol.* 383 (1973) 294–298.

[32] R. D. Allison, E. L. Holmes, J. Nyboer, Volumetric dynamics of respiration as measured by electrical impedance plethysmography, *J. Appl. Physiol.* 19 (1964) 166–173.

[33] G. Brüllmann, K. Fritsch, R. Thurnheer, K. E. Bloch, Respiratory monitoring by inductive plethysmography in unrestrained subjects using position sensor-adjusted calibration, *Respir.* 79 (2010) 112–120.

[34] R. D. Allison, E. L. Holmes, J. Nyboer, Volumetric dynamics of respiration as measured by electrical impedance plethysmography, *J. Appl. Physiol.* 19 (1964) 166–173.

[35] G. Brüllmann, K. Fritsch, R. Thurnheer, K. E. Bloch, Respiratory monitoring by inductive plethysmography in unrestrained subjects using position sensor-adjusted calibration, *Respir.* 79 (2010) 112–120.

[36] J. E. Lewin, An apnoea -alarm mattress, Lancet, 294 (1969) 667–688.

[37] J. Gundersen, K. Dahlin, Monitoring of breathing with a segmental air-filled mattress, *Med. Biol Eng. Comput.* 9 (1971) 541–547.

[38] J. Werthammer, J. Dibenedetto, A.R. Stark, Apnea monitoring by acoustic detection of airflow, *Pediatr.* 71 (1983) 53–55.

[39] A.Jubran, Partitioning of lung and chest-wall mechanics before and after lung-volume-reduction surgery, *Chest* 116 (1999) 1416–1425.

[40] H.K. Walker, W.D. Hall, J.W. Hurst (Eds.), Clinical Methods: The History, Physical and laboratory examinations, Butterworth Publishers; Oxford, 1990, pp. 226–230.

[41] S. Korposh, S-W. Lee, Nano-assembled thin film gas sensors. IV. Mass-sensitive monitoring of humidity using quartz crystal microbalance (QCM) electrodes, *Sens. and Act.* B, 147 (2010) 599-606.

[42] I. Ichinose, T. Kawakami, T. Kunitake, Alternate molecular layers of metal oxides and hydroxyl polymers prepared by the surface sol-gel process, *Adv. Mater.* 10 (1998) 535–539.

[43] K. Ariga, Y. Lvov, T. Kunitake, Assembling alternate dye-polyion molecular films by electrostatic layer-by-layer adsorption, *J. Am. Chem. Soc.* 119 (1997) 2224–2231.

[44] S.-P. Chen, D. Ejeh, D. Levermore, H. Griffith, Y. Bullen, C.M. Hosten, P. Hambright, A Spectroelectrochemical Investigation of the Reduction of the Water-Soluble MnTSPP: SERS Spectra of MnIITSPP, *J. Phys. Chem. B.* 102 (1998) 9101–9108.

[45] Y. Lvov, K. Ariga, I. Ichinose, T. Kunitake, Formation of ultrathin multilayer and hydrated gel from montmorillonite and linear polycations, *Langmuir.* 12 (1996) 3038–3044.

[46] L. Liu, Y. Li, M. Liu, Acidichromism and Supramolecular Chirality of Tetrakis(4-sulfonatophenyl)porphyrin in Organized Molecular Films, *J. Phys. Chem. C.* 112 (2008) 4861–4866.

[47] J. M. Ribo, R. Rubires, Z. El-Hachemi, J.-A. Farrera, L. Campos, G. L. Pakhomov, M. Vendrell, Self-assembly to ordered films of the homoassociate solutions of the tetrasodium salt of 5,10,15,20-tetrakis(4-sulfonatophenyl) porphyrin dihydrochloride, *Materials Science and Engineering* C. 11 (2000) 107–115.

In: Humidity Sensors
Editor: Christopher T. Okada

ISBN 978-1-61209-246-1

Chapter 3

HOLOGRAPHIC HUMIDITY SENSORS

I. Naydenova, J. Raghavendra, S. Martin and V. Toal
Dublin Institute of Technology,
Dublin, Ireland

ABSTRACT

This chapter presents holographic devices for sensing the relative humidity in the environment. The principle of operation of two main types of holographic sensors based on transmission and reflection holograms is described and their properties are theoretically modelled. The effect of different factors such as change of the overall refractive index, change of the hologram's thickness due to swelling/shrinkage of the layer and change in the refractive index modulation due to absorption/desorption of moisture are considered. In the experimental studies emphasis is given to reflection holographic humidity sensors which can provide a visual indication of the relative humidity observed as a change of the colour of the reflected light. The work presented focuses on holograms recorded in acrylamide based photopolymers. The results from studies of their sensitivity, reversibility, relative humidity range of operation and response time are summarised. The operational temperature range is also studied.

The potential use of the humidity sensors as user-verifiable security holograms is described as well as applications in packaging and environmental sensing.

1. Introduction

Holographic sensors for detection of different types of analytes have been attracting significant attention, because they possess a large number of advantages, the most important of which are: possibility for a visually detectable response, no need of expensive instrumentation for sensor interrogation, low cost which allows for the design of disposable sensors, light weight (the sensors are usually made of plastic materials), possibility for miniaturisation and of holographic multiplexing and, with the right choice of materials, the sensors can be easily mass produced. Basically, the holographic sensor consists of a hologram that is disposed throughout a support medium which is analyte sensitive. The process of detection involves a change in the optical characteristics of the hologram caused by the interaction of the analyte with the support medium. The main effort in the development of holographic sensors for visual detection has been focused on reflection holographic sensors [1-9] and on optimisation of the support medium. The optimisation of the support medium targets the achievement of selectivity towards a specific analyte. Reflection holographic sensors sensitive to trypsin [1], water in hydrophobic liquids [2], alcohol levels [3], pH [4], glucose [5,6], l-lactate [7,8] and bacterial spores [9] have been developed. The principle of operation of these holographic sensors [1-9] relies mainly on swelling /shrinking of the hologram support medium when it is exposed to the analyte. Other mechanisms such as changes of the average refractive index and of the refractive index modulation are also possible. Their use in the design of sensors based on transmission holographic gratings has been demonstrated [10]

Recently we have reported a colour changing, reversible, humidity sensitive hologram developed at the Centre for Industrial and Engineering Optics [11-13]. The device has been fabricated in a self developing acrylamide based photopolymer. The advantages of the acrylamide based photopolymer holograms for humidity sensing are their reversibility and the fact that the hologram recording process does not require any wet processing. It was found that the main factor responsible for the colour change in these holograms is the dimensional changes of the polymer film when it is exposed to different humidity levels [13]. The sensitivity of other holographic recording materials to humidity has been known for years [14, 15] and has been used, for example, in silver halide materials to achieve colour control in reflection holograms [14]. Special care is required with other materials [15] to avoid exposure to high humidity as it destroys the recorded holograms. The design of a hologram-based humidity sensor requires a unique combination of properties of the holographic recording material – high sensitivity to humidity and at the same time, reversibility and resistance to damage by humidity. To achieve this combination of properties appears to be very challenging. A humidity sensor using a wavelength-dependent holographic filter with fibre optic links was reported by Spooncer et al [16]. The sensor medium, gelatin, was characterized by a strong hysteresis in humidity response, temperature dependence and variable sensitivity and reported as unsuitable for use as a sensor.

In addition to the reflection holographic sensors, a number of transmission sensors have also been reported. A humidity sensing hologram based on a transmission holographic grating recorded in a modified, polymer-dispersed liquid crystal system with a very fast response time of 350 ms, has been reported [17]. Transmission holographic gratings recorded in zeolite doped photopolymerisable materials have been recently proposed for the development of irreversible humidity sensors [18]. Their principle of operation is based on irreversible change in the refractive index of spatially redistributed hydrophilic nanoparticles which leads to an irreversible change in the refractive index modulation and thus in the diffraction efficiency of the holograms.

2. Background

2.1. Holographic Sensors –Principle of Operation and Types

A holographic sensor is a hologram that is capable of changing its characteristics when it is exposed to the targeted analyte. For example, as a result of detection of the analyte, the sensor could change its diffraction efficiency and/or its spectral response which in visual terms translates into a change of its brightness and/or colour. The hologram is fabricated by exposing a photosensitive material to the optical interference pattern produced when two coherent light beams, usually from a laser, overlap. The material records the variation in light intensity/polarisation as a variation in refractive index, absorption or thickness of the photosensitive layer, and thus a hologram is produced. When at a later stage, the hologram is illuminated with one of the recording beams it will reproduce the other beam which was used at the recording step. Depending on the recording geometry and what optical property of the material is changed during holographic recording, different holograms can be produced and used in the design of a detection system; these are volume or surface relief holograms, transmission or reflection holograms, phase or amplitude holograms.

The holographic humidity sensors reported in this chapter are based on volume phase holograms and a more detailed description of these holograms, their recording, the observable changes in their properties as a result of humidity change and the factors influencing their properties, is presented in the following section.

2.2. Transmission Holograms

2.2.1. Recording and Probing Geometry

A simple transmission volume phase hologram can be recorded when the photosensitive medium is located in the area of overlap of two coherent beams of light of appropriate wavelength. A schematic of the recording geometry is shown in Figure 1a. For simplicity, the two recording beams are assumed to be plane waves. In this case the recorded hologram is a transmission diffraction grating.

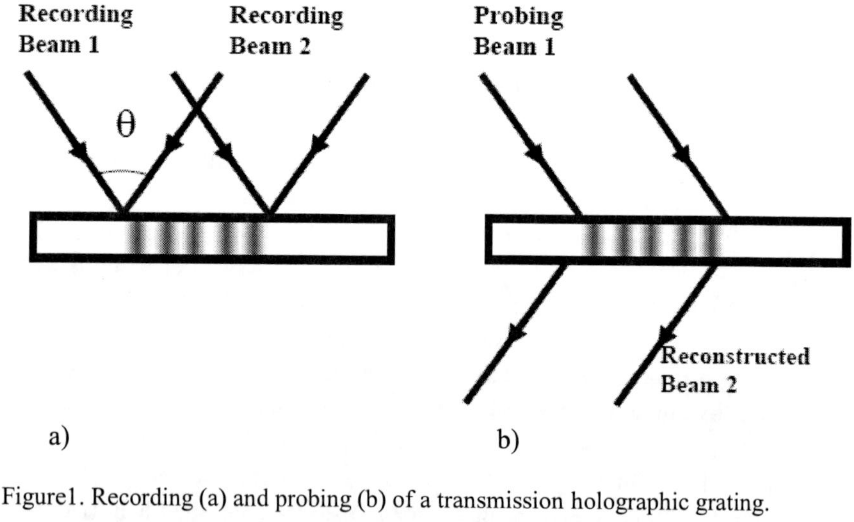

Figure1. Recording (a) and probing (b) of a transmission holographic grating.

The recorded holographic grating will reconstruct one of the recording beams (Fig 1b) when it is probed by the other recording beam. An important parameter of the holographic diffraction grating is its period, Λ, determined by the hlf the angle between the two recording beams ,θ , and it is related to the recording wavelength, λ, by:

$$2\Lambda\sin\theta = \lambda . \qquad (1)$$

The grating period is the inverse of the spatial frequency of recording, which is usually measured in lines/mm. Another important characteristic of such hologram is its diffraction efficiency (DE) which is determined by [19]:

$$\eta = \sin^2\left(\frac{\pi n_1 d}{\lambda\cos\theta}\right) \qquad (2)$$

where n_1 is the refractive index modulation (the difference between the refractive indices in exposed and unexposed areas), d is the thickness of the hologram, λ is the wavelength of the probe beam that is used to measure the diffraction efficiency and θ is the Bragg angle.

2.2.2 Factors Influencing the Properties of the Sensors Based on Transmission Holograms

In order to evaluate the different factors that can influence the sensitivity of the holographic sensor we differentiate (2):

$$\frac{\Delta\eta}{\eta} = \left|\frac{2}{\tan\left(\frac{\pi n_1 d}{\lambda\cos\theta}\right)}\frac{\pi n_1 d}{\lambda\cos\theta}\right|\left(\frac{\Delta n_1}{n_1} + \frac{\Delta d}{d} - \frac{\Delta\lambda}{\lambda} + |\tan\theta|\Delta\theta\right) \qquad (3)$$

It can be seen from this analysis that the two physical parameters of the hologram that are influenced by the presence of the analyte are the refractive index modulation and the thickness of the hologram. The other two parameters, probe wavelength and angle can be controlled externally and they depend on the design of the interrogation system.

Refractive Index Modulation

Although one can anticipate that, in reality, more than one factor can contribute to the diffraction efficiency change, for the sake of simplicity, in the following, it will be assumed that the thickness of the hologram, the probe wavelength and the Bragg angle remain constant. The last two are parameters that can be easily controlled during the interrogation process. We consider a hologram recorded in a photosensitive material whose properties (density, porosity, chemical affinity, charge density, solubility, etc.) differ in the exposed and unexposed regions. If these two regions can adsorb/absorb the analyte to differing extents, an exposure to an appropriate analyte will cause different changes in their refractive indices. Because the exposed and unexposed regions are spatially patterned this will lead to a change in the refractive index modulation Δn_1. Figure 2 shows the dependence of the absolute change of the diffraction efficiency on the initial diffraction efficiency when the relative change of the refractive index modulation is 1%, 5% and 10%. The refractive index modulation, commonly achieved during holographic recording in most photopolymer materials is of the order of 10^{-3}. It can be seen from the graphs that a change in the refractive index modulation from 10^{-5} to 10^{-4} will lead to a change in the diffraction efficiency from slightly less than 1%, up to 9%. Both are easily detectable with the help of instrumentation. Moreover much higher changes are expected [20]. It is seen from the figure that one can optimise the initial diffraction efficiency of the hologram in order to achieve the largest response. The optimum initial diffraction efficiency of the holographic sensor is around 70%.

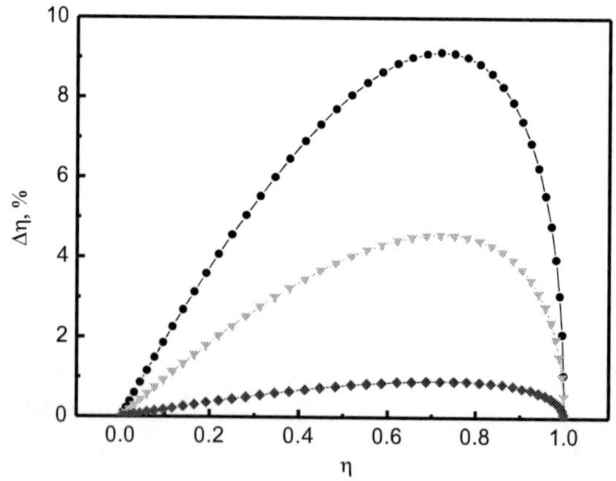

Figure 2. Absolute change of the DE due to a change of the refractive index modulation – 1% diamonds, 5% triangles, 10% circles.

Variation of the Thickness of the Hologram

In the following analysis, for the sake of simplicity we assume that all the other parameters – refractive index modulation change, probe wavelength and angle are fixed. From (3) one can see that thickness change due to shrinking or swelling of the hologram will influence its diffraction efficiency in a manner identical to that shown in Figure 2. The main difference is that normally the changes of the thickness of the hologram are much larger. We have demonstrated [13] that for acrylamide photopolymer hologram thickness changes of order of 10-30% can be readily caused by a change in humidity. In this case a maximum diffraction efficiency change of around 25% is obtainable (Figure 3).

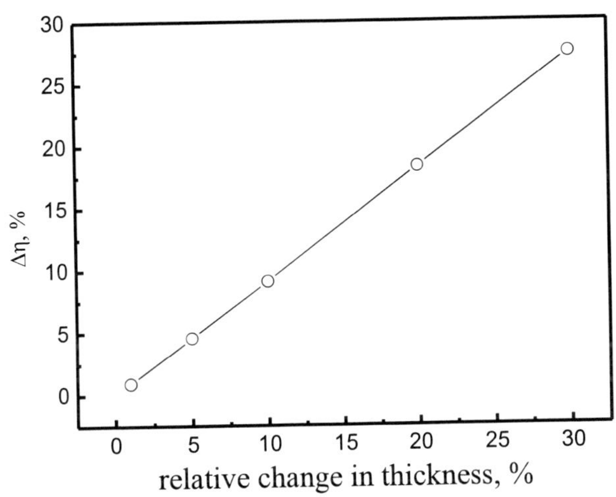

Figure 3. Change of the DE (initial diffraction efficiency is 70%) due to a change of the thickness of the layer caused by the analyte.

Variation of the Probe Beam Wavelength

For most practical applications it will be convenient to keep the probe wavelength constant and thus no wavelength dependent change of the diffraction efficiency will occur.

If a light source with broad spectral range is used to probe a slanted transmission grating then a spectral shift of the maximum DE, caused by shrinking or swelling [21] in the presence of an analyte will be detected. A slanted grating consists of fringes that are slanted at an angle with respect to the layer's surface and any dimensional change of the layer will lead to effective change of the fringe spacing and thus will influence the wavelength at which the DE will peak (see eq.1).

Change in the Bragg Angle

Alternatively, if the probe wavelength is constant but the layer containing the hologram shrinks or swells in the presence of an analyte, a change in the Bragg angle and thus the angular position of the maximum DE is expected in slanted holographic gratings. As one can see from (3) the sensitivity of the sensors in this case can be increased by using higher spatial frequency holographic gratings. No change of the Bragg peak position is expected in unslanted holographic gratings.

2.3. Reflection Holograms

2.3.1. Recording and Probing Geometry

The recording and replay of a reflection hologram can be explained with the help of Figure 4. In this geometry of recording the two coherent beams are incident from the opposite sides of the recording medium. If both light beams are plane waves, the result is a simple diffraction grating whose spatial period depends on the angle between the two beams and is determined by the Bragg' law:

$$\Lambda = \frac{\lambda}{2n_0 \sin\theta}, \tag{4}$$

where θ is the half of the angle between the two recording beam inside the recording medium, λ is the wavelength of the recording light, n_0 is the average refractive index of the recording medium and Λ is the grating period.

When the beams have the same incident angles the interference fringes are parallel to the surface of the recording medium and in this case an unslanted reflection grating is recorded. The incident angles of the two recording beams could also be different and then the fringes would be at an angle with respect to the recording medium surface and the recorded grating would be slanted.

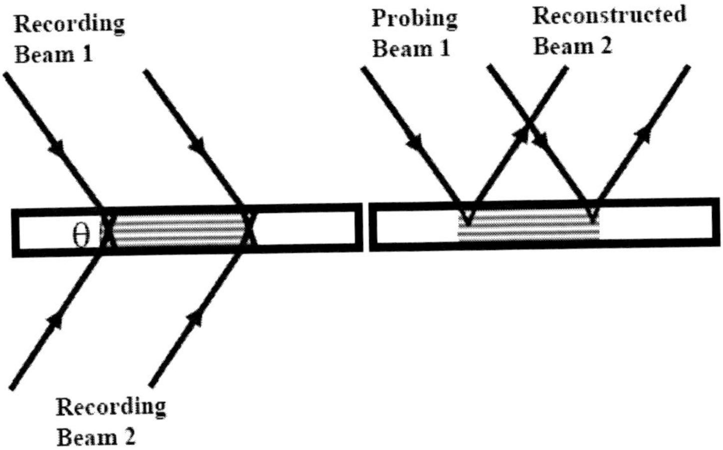

Figure 4. Recording (a) and probing (b) of a reflection holographic grating.

2.3.2. Factors Influencing the Properties of the Sensors

The property that most conveniently can be observed to change in reflection holograms in the presence of an analyte is the wavelength of the diffracted light. When illuminated with a light with a broad spectral range, a reflection hologram will selectively diffract and the maximum diffraction efficiency will occur at a wavelength which satisfies eq. (4):

$$\lambda = 2n_0 \Lambda \sin\theta \quad (4)$$

In order to determine the factors influencing the change in the spectral peak of the reflected light, we can differentiate eq.4.

$$\frac{\Delta\lambda}{\lambda} = \frac{\Delta n_0}{n_0} + \frac{\Delta\Lambda}{\Lambda} + \cot\theta.\Delta\theta \quad (5),$$

where $\Delta\lambda$, Δn_0, $\Delta\Lambda$ and $\Delta\theta$ are the changes in wavelength, average refractive index, grating period and the Bragg angle respectively.

Average Refractive Index Change

The influence of the change in the average refractive index of the hologram on the diffracted light peak wavelength is shown in Figure 5.

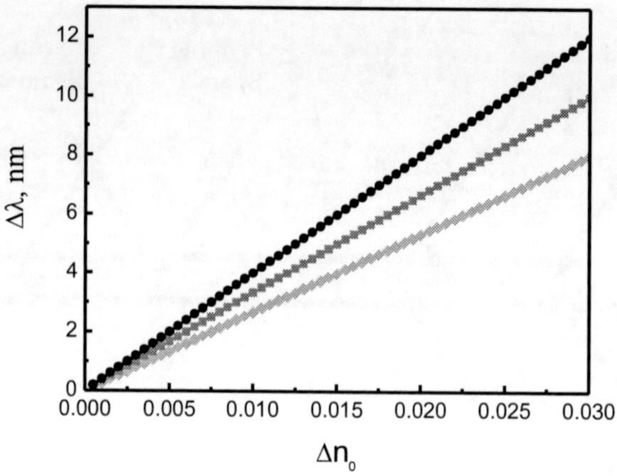

Figure 5. Dependence of the peak wavelength change on the average refractive index change. The initial wavelength is 600nm (black circles), 500 nm (dark grey squares), 400nm (light grey diamonds). The initial average refractive index is assumed to be 1.5.

It can be seen from Figure 5 that in order to obtain a larger change in the peak wavelength, with the grating period and probe angle remaining constant, it would be beneficial to have the hologram recorded at a longer wavelength. The initial average refractive index in these simulations was 1.5. The absolute change in the peak wavelength can be increased by choosing materials with lower average refractive index, which should be possible, bearing in mind that for some larger size analytes, materials with larger pores would be required.

Grating Period Change

As one can see from eq.5, the change in the peak wavelength also depends on the change in the spatial period which in turn depends on the relative change in the dimensions of the hologram, caused by swelling or shrinkage. A simulation assuming that the average refractive index and probe angle remain constant, reveals that changes in dimensions could produce a much more easily observable change in the peak wavelength (Figure6). A practically achievable dimensional change of 30 % would produce above 100 nm shift depending on the initial peak wavelength. In this simulation an unslanted reflection hologram is considered and the relative change in hologram thickness is equal to the relative change in grating period.

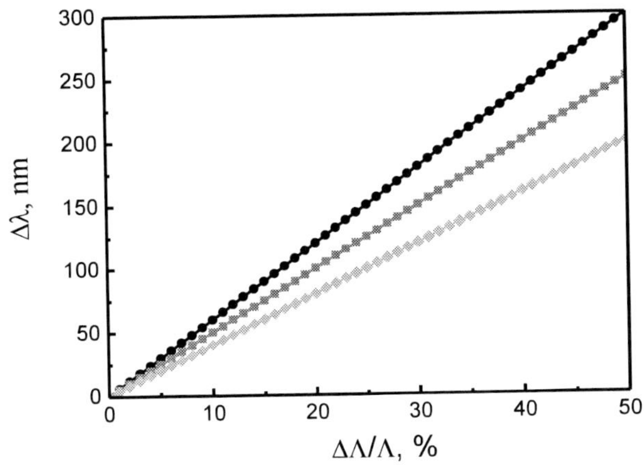

Figure 6. Dependence of the peak wavelength change on the relative dimensional change. The initial wavelength is 600nm (black circles), 500 nm (dark grey squares), 400nm (light grey diamonds).

Angle of Observation

It can be seen from eq.(4) that different wavelengths/colours can be observed at different probe angles even without the presence of an analyte. That is why, in order to ensure that the introduced change is solely due to a change in the physical parameters of the holographic sensor, it is necessary either to fix the probe angle or to make sure that the thickness of the hologram is large enough so that no colours other than the original one are reconstructed and easily observable.

3. Reversible Holographic Humidity Sensor Based on a Reflection Hologram

3.1. Preparation and Methods for Characterisation of the Holograms

3.1.1. Photopolymer Layer Preparation

The photopolymer material used in this study is a self-processing acrylamide based photopolymer developed at the Centre for Industrial and Engineering Optics, Dublin Institute of Technology for recording in reflection mode [11]. The photosensitive layer consists of two monomers - acrylamide and N,N Methylene bisacrylamide, a polymeric binder - polyvinyl alcohol, a photosensitiser-Erythrosin B and an initiator triethanolamine (TEA). Different mixtures of these components were studied in order to obtain optimum diffraction efficiency and optimum humidity response of the reflection holographic optical element. The preparation of one of the formulations that has been studied, is described below. Two ml of triethanolamine was added to 17.5 ml stock solution of polyvinyl alcohol (PVA) (10% w/v). Then the monomers, 0.6 g acrylamide and the crosslinker, 0.2 g of N,N Methylene bisacrylamide, were added. Finally, 4 ml of Erythrosin B dye was added from a stock solution with concentration of 1.1 mM. 0.5 ml of this solution was spread on a levelled 25x75 mm^2 plate. The samples were dried for 5-10 hours. When the dependence of the humidity response on the hologram thickness was studied, the layers were prepared from different amounts of photopolymer solution varying from 0.15 to 0.75 ml per 25x75 mm^2 slide.

The thickness of the dry layers was measured by cutting the photopolymer layer using a sharp scalpel and measuring the vertical dimension of the cut with a white light interferometric (WLI) surface profiler ***MicroXAM S/N 8038***.

3.1.2. Recording of the Holographic Optical Element

The reflection gratings were recorded using recording geometry described previously [13]. The frequency doubled output from a $NdYVO_4$ laser (Verdi 05) was used to record the holographic gratings at wavelength of λ=532 nm. The total exposure of recording did not exceed 100 mJ/cm^2. Some of the humidity indicators incorporating a holographic image of an object were recorded as Denisyuk type holograms.

3.1.3. Measurement of the Spectral Response

A controlled environment chamber with humidity and temperature control system was purchased from Electro-Tech Systems, Inc. The system model 5503-11 consists of a glove box, a microprocessor controller, a dehumidification system, an ultrasonic humidification system and thermoelectric cooling and heating systems. The relative humidity in the chamber can be maintained to better than ± 1 % RH of the set point. The operational range is from 5 % to 100 % RH. A fast response capacitive type sensor providing measurement accuracy of ± 2 % RH over the range of 0-90 % and 3 % from 90 to 100 % RH is incorporated in the chamber. The temperature in the chamber is maintained within 0.1 °C of the set point.

The optical test set-up was assembled in the humidity chamber. The light from an AvaLight-HAL-S light source was used to probe the reflection holograms. In order to guide the light into the humidity chamber to the hologram, the light was coupled to a fibre and then collimated by a lens. The light diffracted by the probed hologram was then coupled to another fibre by means of a second lens and guided to a spectral analyser Model AvaSpec-2048. In order to obtain the calibration curve for a specific device the relationship between the wavelength corresponding to the maximum intensity in the spectral response of the hologram and the relative humidity in the chamber was determined.

When the visual colour change of the sample was studied, the hologram was illuminated by a white light spot lamp and the holographic image was recorded by a digital camera model Sony Cyber-shot DSC300.

3.1.4. Measurement of the Response Time

Measurement of the response time of the humidity indicator was carried out by recording the change of the wavelength position of the maximum intensity in the spectral response of the hologram at different times after achieving preset humidity. The increment in humidity was 10%. The time for achieving the preset humidity was less than 1min. The change in the peak position during this latter time interval for layers with thickness above 30 μm was negligible. Only the changes in peak position, after achieving the preset humidity were recorded.

3.2. Range of Operation of the Humidity Sensor

3.2.1. Characterisation of the Range of Operation

The typical spectral response of a reflection holographic grating recorded in a 30 μm thick photopolymer layer is presented in Figure 7. As expected from such a holographic optical element [19] it is very selective with respect to the wavelength of the diffracted light - the full width at the half maximum is only 2.8 nm. This explains the single colour appearance of the reconstructed image from a reflection hologram in Figure 8. The position of the peak in the spectral response curve can be determined from these data and its change with the change of the relative humidity in the controlled environment can be followed in order to calibrate the humidity indicator.

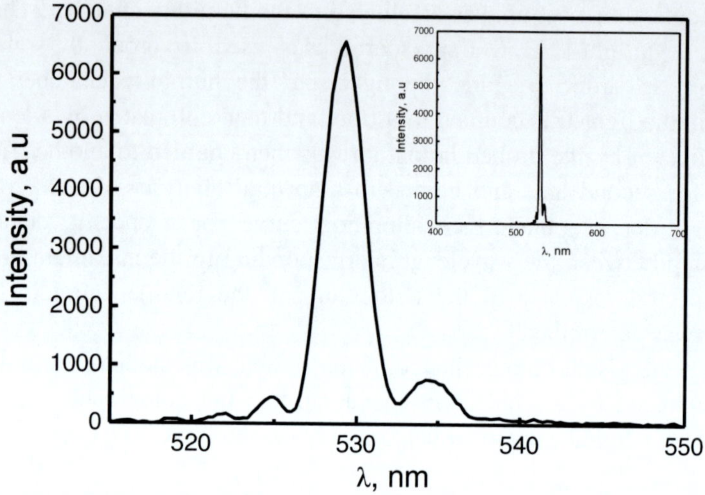

Figure 7 Spectral response of reflection holographic grating recorded in 30 μm thick layer. Copyright 2007 Society of Photo-Optical Instrumentation Engineers (ref. 11).

Figure 8. Change of the image colour when a hologram is exposed to different humidity levels.

An example of the humidity response of a reflection grating recorded in a photopolymer layer is shown in Figure 9. It is seen that when the relative humidity changes from 12 % to 80 % the position of the peak in the spectral response also changes by 168 nm. With increasing humidity the position of the spectral peak shifts towards the red region of the spectrum. Such a large shift in the peak wavelength can be readily explained by the swelling of the photopolymer layer and the consequent increase of the recorded fringe spacing. When the relative humidity decreases, the spectral peak shifts towards the blue region of the spectrum and this can be explained by shrinkage of the photopolymer layer and decrease of the recorded fringe spacing. It should be noted that the relation between the position of the spectral peak and the value of the relative humidity is nonlinear. The humidity indicator is more sensitive at higher than at lower relative humidity.

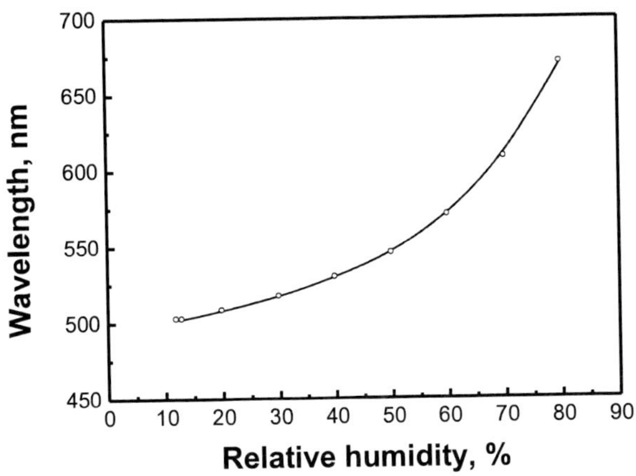

Figure 9.Humidity response of a reflection grating recorded in 30 μm thick photopolymer layer.

The relative humidity during the recording of the grating was 29 %.

3.2.2. Controlling the Range of Operation

Thermal Treatment of the Layers at Elevated Temperatures

In order to explore different methods for modification of the operational range of the humidity indicator we studied the effect of exposure to elevated temperature on previously recorded humidity sensors. An example of the influence of the elevated temperature on the humidity response of the sensors is presented in Figure 10. It is seen that the heating of the sample causes a shift towards lower wavelengths in the calibration curve. This is most probably due to irreversible shrinkage. The shift in the calibration curve can be used to adjust the exact colour of the hologram at a specific humidity. It can also be used to make the holographic image invisible at a specific humidity, as the image will be reconstructed in the UV spectral range, and appear only when the hologram is exposed to a higher humidity level.

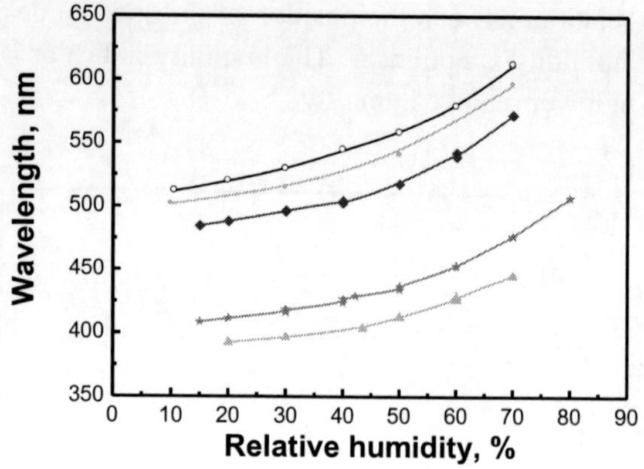

Figure 10. Influence of exposure to elevated temperatures on the humidity response of the sensors: before heating (open circles), after heating to 100 °C (light gray squares), 110 °C (dark grey diamonds), 125 °C (grey stars); 145 °C (light grey triangles).

Variation of the Humidity at the Recording Step

It was observed that the relative humidity in the laboratory during the recording process is important for the final colour appearance of the hologram and also influences the calibration curve. For instance two different humidity sensors were recorded at relative humidity values of 38% and 29%. The calibration curves of the two sensors are shown in Figure 11.

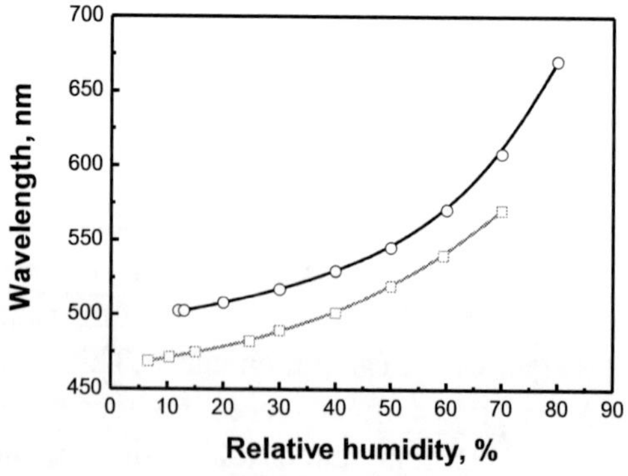

Figure 11. Calibration curves for sensors recorded at 29 % RH (black) and 38 % RH (grey).

It can be seen from the figure that both calibration curves are shifted so that the recording wavelength of 532 nm is observed at higher humidity than the humidity at recording. This implies that the photopolymer layers have shrunk during or immediately after the recording and, when exposed to the same humidity as at recording, the replayed wavelength is blue shifted. In order to replay the original wavelength one needs to expose the layers to higher humidity than the humidity at recording. The results also show that by varying the relative humidity at the recording stage one can modify the calibration curve of the sensor.

3.3. Temperature Dependence

In order to evaluate the performance of the sensor as a relative humidity sensor it was important to study its temperature response. The temperature range in this study was restricted by the operating range of the environmental chamber. It was observed that there was little or no temperature dependence of the peak wavelength for temperatures up to 30 °C. Above this temperature the sensor becomes temperature sensitive and its response needs to be recalibrated (Figure 12). A detailed study of the temperature dependence of the hologram is presented in [13].

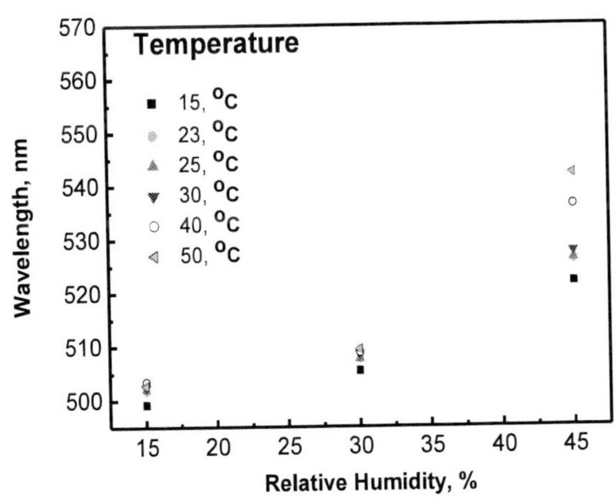

Figure 12. Temperature dependence of the humidity response of a holographic sensor recorded in 30 μm thick photopolymer layer.

3.4. Response Time

3.4.1. Characterisation of the Response Time

The response of the sensor was characterised by measuring the peak wavelength position with time after a fast change in the relative humidity. The results shown in Figure 13 reveal that less than two minutes were required for a hologram which was recorded in 30 μm thick photopolymer layer to achieve stable response. After that no further change in the peak wavelength was observed. The actual change in the spectral characteristic started immediately after the change in humidity was introduced. Approximately the same time was required in three separate experiments when the relative humidity was increased by 10 % , from 30 to 40 %, from 40 to 50 % and from 50 to 60 %.

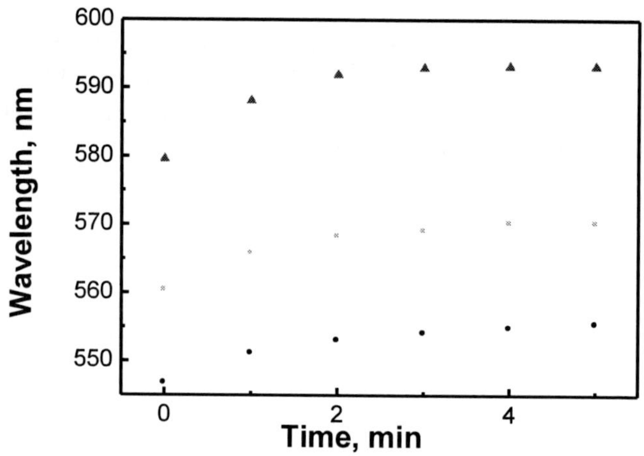

Figure 13. Change of the peak wavelength with time after the relative humidity was changed from 30 to 40 % (black circles), from 40 to 50 % (gray squares) and from 50 to 60 % (dark grey triangles).

The response time was also evaluated in the case when the humidity was decreased. An example of these results is shown in Figure 14. A hologram, which initially reconstructs the image in blue colour, was exposed to high relative humidity (close to 100 %) by breathing on it. The time required for the hologram to recover its initial dimensions and to reconstruct the image in the original colour can be estimated from the pictures. It is seen that it takes approximately two minutes for complete recovery of the initial colour.

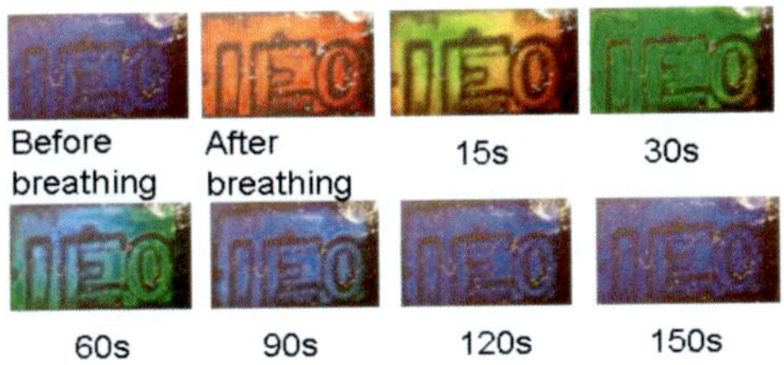

Figure14. Complete recovery of the original colour after introducing a red shift by breathing on the hologram is observed within 120 s. The thickness of the holograms was 30 μm.

3.4.2. Control of the Response Time of the Sensor

The response time of layers of different thickness was studied. It was observed that the response time of the hologram can be tuned by varying the thickness of the layers [11]. Layers of thickness 30 and 90 μm are compared in Figure15. The change in the peak wavelength with respect to the initial value at 50 % relative humidity was studied after a fast change in the relative humidity in the environmental chamber from 50 to 60 %. It is seen that the thicker layers require longer time to reach the final value of the peak wavelength. The shift is normalised to the maximum wavelength shift which was 17.6 and 20 nm in 30 μm and 90 μm thick layers respectively. The dependence of the response time on the layer thickness shows potential for flexibility in the design of the humidity sensors when different response times are required.

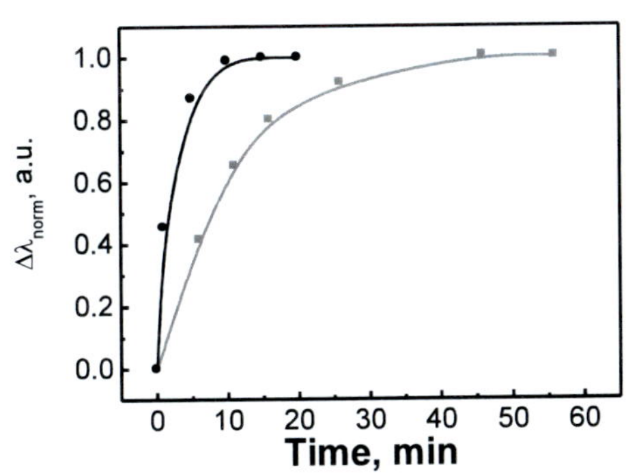

Figure15. Time response of holograms recorded in 30 μm (black circles) and 90 μm (grey squares) thick layers.

3.5. Reversibility

Another important aspect of the behaviour of the humidity indicator is its reversibility which was studied by exposing the hologram to a number of cycles of humidity change. Pictures of a hologram exposed to 20 % and 80 % relative humidity in seven consecutive cycles are shown in Figure 16. As can be seen from the figure the acrylamide based holographic sensor is fully reversible. Although at high humidity some moisture could be deposited on the surface of the hologram, this did not damage the hologram and, when it was brought back to low humidity, the moisture evaporated and the observed holographic image was unchanged.

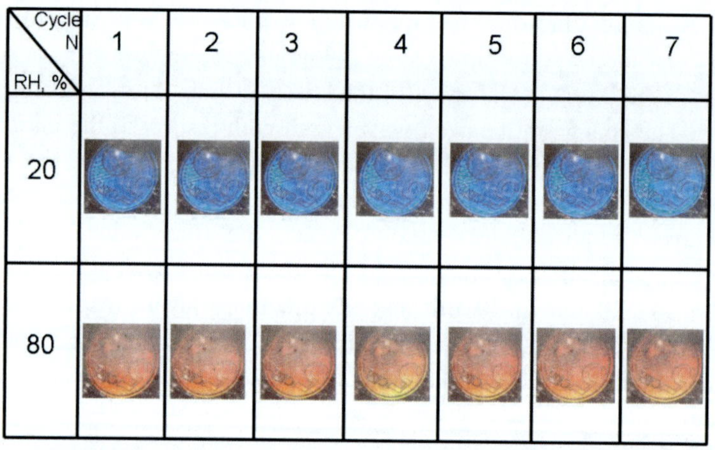

Figure 16 Reversible response of the holographic humidity sensor

3.7.1. Characterisation of the Viewing Angle

A bright (high diffraction efficiency) hologram which changes colour when the environmental humidity changes is not sufficient for most commercial applications, unless the observer can see it over a wide range of angles. The holographic humidity indicator should be designed in such a way that the angular field of view of the indicator is large. The field of view of the hologram depends on different parameters, the most important of which is the thickness of the photosensitive layer in which the hologram is recorded.

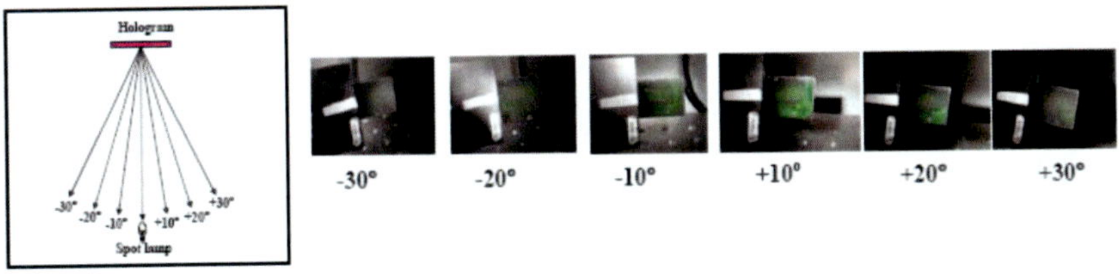

a b

Figure 17 a) Experimental set-up used in viewing angle studies and b) photographs of the reconstructed image from a reflection hologram.

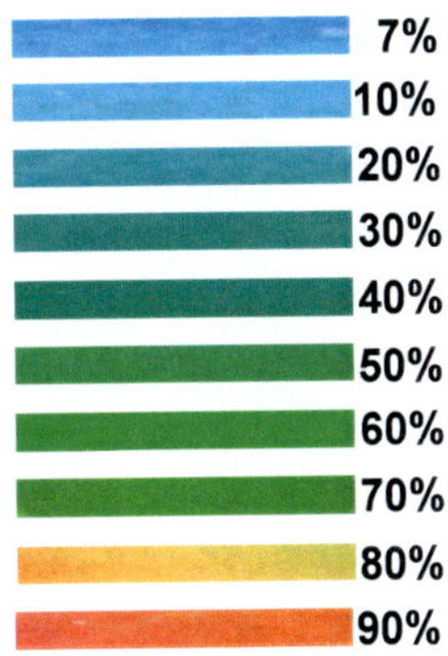

Figure 18. Colour coded relative humidity obtained from the response of a holographic sensor.

Another parameter is the nature of the object that is used to record the hologram. A hologram recorded using a diffused object beam would have larger field of view. The field of view of a reflection hologram recorded in a photosensitive layer of thickness 28 μm was studied using the experimental set-up shown in Figure17a. A spot lamp was used to illuminate the hologram and the photographs of the reconstructed image were taken from different angles ranging from - 30° to + 30°. The results are shown in Figure17b. The hologram can be viewed from a wide angle and the field of view of the hologram is 60°.

3.7.2. Colour Calibration of the Sensor

In order to calibrate the holographic humidity sensor, reflection gratings were recorded and then reconstructed. The reconstructed image was photographed at relative humidities ranging from 7 to 90 %. Then the colours were used to generate the chart shown in Figure18. It can be seen that the colours are well distinguished and the sensor can be used for low cost indication of the relative humidity in the environment.

5. Potential Applications of Holographic Humidity Sensors

Holographic humidity sensors can have applications in a wide range of industries from brand protection/security to health and environmental monitoring and packaging. Their field of application can be broadly divided into two categories - those in which they are used quantitatively as a sensor of humidity and those in which their colour-changing attributes are exploited as a security / authentication device.

5.1. Humidity Sensing:

In this application the key advantages to be exploited are the low cost, light weight and compactness and the fact that the sensor requires no external power source (other than ambient lighting) in order to be read. This reduces cost very considerably when compared to other humidity sensors and means that the sensor can be incorporated into packaging for use with electronics, pharmaceuticals, food and any other product where humidity in the internal environment of the package is important. Costs are low enough to allow single use, disposable sensor cards to rival the humidity sensor cards produced by Sud-Chemie and other companies. These products change from blue to pink as the humidity changes. The holographic alternative has the advantages of providing a whole range of colours (with a corresponding humidity value) in each spot, the potential for images and logos including 3D images and, importantly, the ability to display holographic numbers or text when a critical humidity is passed. This is achieved by multiplexing more than one hologram, each visible at a different humidity value.

There is also the potential for devices built around holographic sensors. A cheap light source and spectrometer convert the hologram into a highly accurate sensor for humidity.

5.2. Security Holograms, Authentication

In this application, the ability to respond quickly to humidity change is a key issue because the hologram is used as a security hologram with the unique attribute of being able to change colour when breathed on. Colour change must be rapid and highly visible but remain at the new colour for a suitable length of time (ideally more than five seconds). A change of image/text on breathing is also possible, again by multiplexing holograms.

A company called Smart Holograms based in the UK produces a silver halide –based product that can be used as a user-verifiable security hologram www.smartholograms.com/. According to a February 2010 press release from Smart Holograms "Verif-EYE™ shows a visual holographic image/color that transforms into a different image/color upon detection of human breath or a drop of water depending on how the sensor is programmed."

IEO's colour changing hologram [22] can also be used in security applications and has the advantage of being made in a self-developing low cost material, simplifying the mass production process, while still retaining the high quality 3D image characteristics of volume holograms.

6. Conclusion

An acrylamide based photopolymer can be used to fabricate relative humidity sensors based on reflection holograms. The response of the sensors can be controlled by varying the composition of the recording photopolymer material, the thickness of the holograms, the humidity at the recording stage and by exposure to elevated temperatures. The sensors are reversible and can visually indicate the relative humidity in the environment by a change in the colour of the reconstructed holographic image.

7. Acknowledgments

The authors acknowledge the financial support of Enterprise Ireland, ULYSSES Travel Grant 2010 and the SARA Programme at Dublin Institute of Technology. Brendan Ring is acknowledged for his support as commercialisation manager. Katie Kelly and Hoda Akbari are acknowledged for their technical support.

8. REFERENCES

[1] R. B. Millington, A. G. Mayes, J. Blyth, and C. R. Lowe, "A Holographic Sensor for Proteases", *Anal. Chem.*, 67, 4229-4233, (1995)

[2] J. Blyth, R. B. Millington, A. G. Mayes, E. R. Frears and C. R. Lowe, "Holographic Sensor for Water in Solvents", *Anal. Chem.*, 68, 1089-1094, (1996)

[3] A. G. Mayes, J. Blyth, M. Kyrölåinen-Reay, R. B. Millington, and

[4] C. R. Lowe, "A Holographic Alcohol Sensor", *Anal. Chem.*, 71, 3390-3396, (1997)

[5] A. J. Marshall, J. Blyth, C. A. B. Davidson, and C. R. Lowe, "pH-Sensitive Holographic Sensors", *Anal. Chem.*, 75, 4423-4431, (2003)

[6] X. Yang, M.-C. Lee, F. Sartain, X. Pan, and C. R. Lowe, "Designed Boronate Ligands for Glucose-Selective Holographic Sensors", *Chem. Eur. J.*, 12, 8491 – 8497 (2006)

[7] X. Yang, X. Pan, J. Blyth, C. R. Lowe, "Towards the real-time monitoring of glucose in tear fluid: holographic glucose sensors with reduced interference from lactate and pH", *Biosensors and Bioelectronics*, 23, 899–905, (2008)

[8] X. Yang, M.-C. Lee, F. Sartain, X. Pan, and C. R. Lowe, "Designed Boronate Ligands for Glucose-Selective Holographic Sensors", *Chem. Eur. J.*, 12, 8491 – 8497, (2006)

[9] F. K. Sartain, X. Yang, and C. R. Lowe, "Complexation of l-Lactate with Boronic Acids: A Solution and Holographic Analysis", *Chem. Eur. J.*, 14, 4060 – 4067, (2008)

[10] D. Bhatta, G. Christie, B. Madrigal-Gonzalez, J. Blyth, C.R. Lowe, "Holographic sensors for the detection of bacterial spores", *Biosensors and Bioelectronics*, 23, 520–527, (2007)

[11] C. Ley, I. Calderara and D. J. Lougnot, "Holographic gratings recorded in polymer hydrogels—an original application as a sensor in aqueous environment", *Meas. Sci. Technol.*, 8, 997–1000, (1997)

[12] I. Naydenova, R. Jallapuram, V. Toal and S. Martin, "Hologram-based humidity indicator for domestic and packaging applications", SPIE proceedings, V 6528, 652811, (2007)

[13] I. Naydenova, R. Jallapuram, V. Toal, S. Martin, " A visual indication of environmental humidity using a colour changing hologram recorded in a self-developing photopolymer", *Appl. Phys. Lett.*, 92, 031109 (2008)

[14] I. Naydenova , J. Raghavendra, V. Toal and S. Martin, "Characterisation of the humidity and temperature responses of a reflection hologram

recorded in an acrylamide-based photopolymer", *Sensors and Actuators B,: Chemical*, 139, 35-8, (2009)

[15] D. R. Wuest and R. S. Lakes, "Color control in reflection holograms by humidity" *Applied Optics*, 30, 2363-7, (1991)

[16] G. Naik, A. Mathur, and S. Pappu, "Dichromated gelatin holograms: an investigation of their environmental stability", *Appl. Opt.*, 29, 5292-5297, (1990)

[17] R. C. Spooncer, F. A. Al-Ramadhan and B. E. Jones, "A humidity sensor using a wavelength-dependent holographic filter with fibre optic links", *International Journal of Optoelectronics,* 7, 449-52, (1992)

[18] J. Shi, V. K. S. Hsiao and T. J. Huang, "Nanoporous polymeric transmission gratings for high-speed humidity sensing", *Nanotechnology*, 18, 465501, (2007)

[19] E. Leite, Tz. Babeva, E.-P. Ng, V. Toal, S.Minatova and I. Naydenova, "Optical Properties of Photopolymer Layers Doped with Aluminophosphate Nanocrystals", *J. Phys. Chem. C*, *114* (39), 16767–16775, (2010)

[20] H. Kogelnik, ''Coupled wave theory'',*Bell Syst. Tech. J.*, 48, 2909 (1969)

[21] E. Leite, I. Naydenova, S. Mintova, L. Leclerq and V. Toal, "Photopolymerisable nanocomposites for holographic recording and sensor application", *Applied Optics*, 49, 3652-60, (2010)

[22] Gallo J. T. and Verber C. M., "Model for the effects of material shrinkage on volume holograms", *Applied Optics*, 33, 6797-6804, (1994)

[23] I. Naydenova, S. Martin, H. Sherif, J. Raghavendra and V. Toal, "A holographic sensor", PCT/IE2006/000134 (WO 2007/060648).

In: Humidity Sensors
Editor: Christopher T. Okada

ISBN 978-1-61209-246-1

Chapter 4

SURFACE MODIFIED ELECTROSPUN NANOFIBROUS MEMBRANES FOR HUMIDITY DETECTION BASED ON QUARTZ CRYSTAL MICROBALANCE TECHNIQUE

***Xianfeng Wang*[1,2,3], *Bin Ding*[1,2,1], *Jianyong Yu*[2] *and Gang Sun*[4]**

[1] State Key Laboratory for Modification of Chemical Fibers and Polymer Materials, Donghua University, Shanghai 201620, China

[2] Nanomaterials Research Center, Modern Textile Institute, Donghua University, Shanghai 200051, China

[3] College of Textiles, Donghua University, Shanghai 201620, China

[4] Fibers and Polymer Science, University of California, Davis, CA 95616, USA

ABSTRACT

Electrospinning exhibits the unique ability to produce diverse forms of fibrous assemblies with remarkable specific surface area and high porosity, making electrospun nanomaterials highly attractive to ultra-sensitive sensors and other nanotechnological applications. In this chapter, electrospun fibrous chitosan (CS) membranes modified with

1 *E-mail: binding@dhu.edu.cn.

polyelectrolytes (polyacrylic acid (PAA) and polyethyleneimine (PEI)) were used as sensing materials coated on quartz crystal microbalance (QCM) for humidity detection. A series of nanofibrous membranes containing polyelectrolytes are fabricated on QCM and characterized regarding their morphology, sensitivity and hysteresis. Sensing experiments are carried out by measuring the resonance frequency shifts of QCM due to the additional mass loading. The results indicated that the response of the sensors increased by three to four orders of magnitude with increasing the relative humidity (RH) from 8 to 95% at room temperature, exhibiting high sensitivity and that, in the range of 20-95% RH, the $Log(\Delta f)$ showed a good linearity. Additionally, the prepared sensors exhibited remarkably enhanced sensitivity with increasing the coating loads of polyelectrolytes on QCM. Furthermore, the dynamic Langmuir adsorption model was used to determine the kinetic parameters such as adsorption rates and Gibbs free energy for various RH between 70 and 95%.

INTRODUCTION

Humidity control and detection are of great interest to a wide area, such as agriculture, meteorology, clinical, and manufacturing [1-2]. Up to now, various sensing technologies, including impedance, capacity, optics and quartz crystal microbalance (QCM), have been adopted to detect humidity [3-4]. Among these sensing approaches, QCM, one of a broad class of acoustic wave techniques, whose history can dates back to 1880 when Pierre and Jacques Curie discovered the piezoelectric effect, where pressure exerted on a small piece of quartz causes an electrical potential difference between the deformed surfaces. The essence of QCM as an analytical tool lies in its ability to detect small mass changes of even smaller than nanogram order deposited on crystal surface and act as a sensor. Interfacial mass changes can be related to changes in the QCM oscillation frequency by applying Sauerbrey's equation [5]:

$$\Delta f = -2 f_o^{\,2} \Delta m / A(\mu\rho)^{1/2} \tag{1}$$

where Δf is the measured frequency shift, f_o is the fundamental frequency of a bare QCM chip, Δm is the mass change per unit area, A is the electrode area, ρ is the density of quartz, and μ is the shear modulus of quartz crystal.

To date, a number of materials such as ceramics, polymers, self-assembled monolayers, dendrimers, and waxes, have been prepared as sensitive coatings on QCM to improve the sensor sensitivity and selectivity for vapor and humidity [6]. It is well known that the sensitivity of QCM sensors towards a specific analyte is enhanced by increasing the specific surface area of the sensing materials. Therefore, recent efforts have been put in the exploration of a humidity sensor based on nanostructured materials, taking advantage of their large surface areas, and hence high surface activity and gas sensitivity [7].

Many approaches for fabricating humidity-sensitive membranes, such as dip coating [8], spin coating [9], layer-by-layer self-assembly [10], and ink-jet printing [11], have been reported. Okur et al. [12] deposited water soluble calixarene derivative films on the electrode of QCM to detect humidity between 29% and 78%. The results of sensing experiments indicated that more work should be carried out to further improve the humidity sensitivity. Electrospinning is a highly versatile technique to process solutions or melts, mainly of polymers, into continuous fibers with diameters ranging from a few micrometers to a few nanometers. Electrospinning exhibits the unique ability to produce diverse forms of fibrous assemblies [13-15]. The remarkable specific surface area and high porosity make electrospun nanomaterials highly attractive to ultrasensitive sensors and other nanotechnological applications [16-17]. Therefore, the electrospun fibrous membranes with large specific surface used as sensing materials or templates would be an ideal candidate to replace the flat films.

In the current work, the fibrous CS membranes as the template, taking advantage of their large surface areas, were firstly deposited on QCM via electrospinning. And then PAA and PEI solutions were dropping casted onto CS fibrous membranes, through which PAA-CS and PEI-CS membranes were obtained. The resultant composite membranes were studied as sensitive coatings for humidity detection and the sensing characteristics such as response, hysteresis, and linearity were discussed. In order to investigate the effect of coating load of sensing membranes on sensor sensitivity, various coating loads PEI-CS membranes were prepared. The Langmuir model was used to analyse the adsorption and desorption kinetic of the PEI-CS membranes between 70% and 95% RH.

EXPERIMENTAL

Preparation of Polymer Solutions

The starting materials included chitosan (CS, M_w 2.1×10^5 KD, DD 92%, Yuhuan Ocean Biochemical Co., China), polyacrylic acid (PAA) (M_w 250,000, Wako Pure Chemical Industries, Ltd., Japan) and polyethyleneimine (PEI) (M_w 70,000, Alfa Aesar Co., Ltd.) aqueous solution of 30 wt%. The chemical structures of CS, PAA and PEI are shown in scheme 1. Pure water with a resistance of 18.2 MΩ, ethanol (99.7 wt%, Shanghai Zhenxing Chemical Co., China), and acetic acid (99.5 wt%, Shanghai Lingfeng Chemical Reagent Co., Ltd., China) were used as solvents. The acetic acid solution of 90 wt% was prepared by diluting acetic acid solution of 99.5 wt% with pure water. A 2% CS solution for electrospinning was prepared from CS powder and 90% acetic acid solution with vigorous stirring. PAA (8 wt‰) and PEI (8 wt‰) solutions for drop casting were prepared in the cosolvent of ethanol and water, the weight ratio of ethanol to water was 2/1.

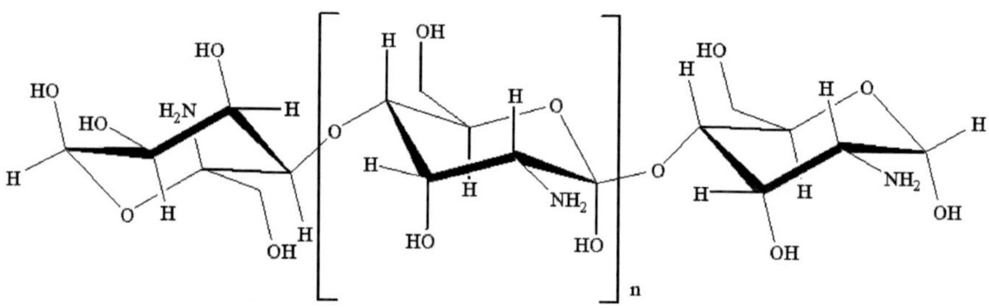

Chitosan, CS

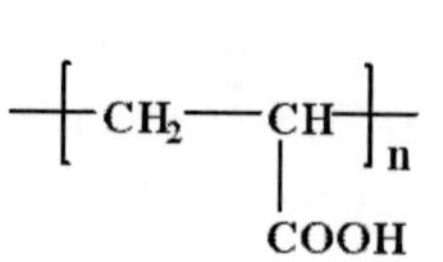

Polyacrylic acid, PAA

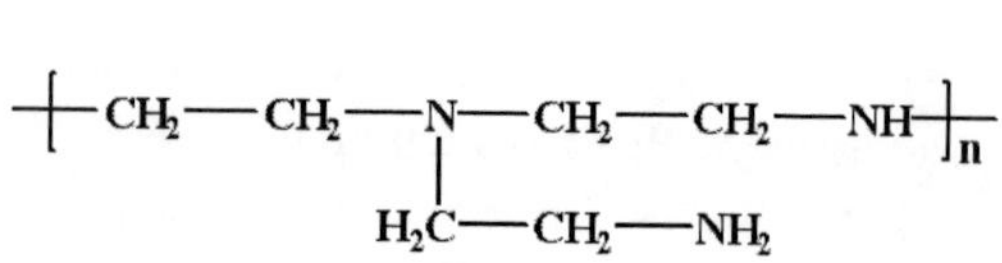

Polyethyleneimine, PEI

Scheme 1. Chemical structures of CS, PAA and PEI.

Fabrication of Sensing Membranes on QCM

Figure 1a shows the schematic of fibrous membranes deposition on the electrode of QCM via electrospinning. The CS solutions were placed in a syringe that was connected to the positive electrode of a high voltage power supply (DW-P303-1ACD8, Tianjin Dongwen High Voltage Co., China). Electrospinning of the CS solutions was conducted at a 0.3 mL/h feeding rate using a syringe pump (LSP02-1B, Baoding Longer Precision Pump Co., Ltd., China) and a 15 kV applied voltage. The fibrous membranes were deposited on the grounded electrode of QCM (5 MHz, AT-cut quartz crystal with Au electrodes) at a 15 cm tip-to-collector distance until the required frequency shift was obtained. The coating loads of CS fibrous membranes on QCM were regulated at about 1200, 4000, and 5000 Hz, respectively. The resonance frequencies were measured by a QCM digital controller (QCM200, Stanford Research Systems). Fibrous membranes coated QCMs were dried at 25 ^{o}C in vacuum for 30 min prior to the subsequent modifications and characterizations. The PAA-CS and PEI-CS composite membranes were prepared by dropping casted PAA (8 wt‰) and PEI (8 wt‰) solutions onto CS fibrous membranes (Figure 1b).

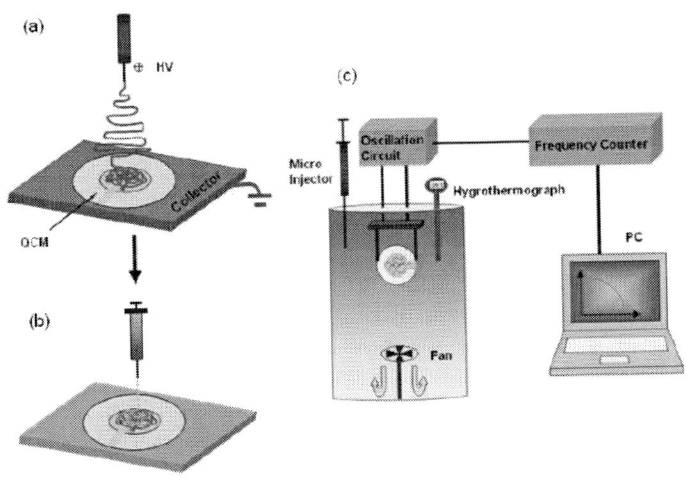

Figure 1. (a) Schematic diagram illustrating the deposition of fibrous membranes onto the electrode of QCM via electrospinning. (b) Schematic diagram illustrating the deposition of polyelectrolyte solutions onto the fibrous membranes via drop casting method. (c) Schematic diagram of a relative humidity testing system.

After the solvents completely evaporated at room temperature, the membranes coated QCM sensors were dried at 25 °C for 2 h in vacuum.

The morphology of membranes on QCM was examined with field emission scanning electron microscopy (FE-SEM) (S-4800, Hitachi Ltd., Japan). The diameters of fibers were measured by using image analyzer (Adobe Photoshop CS2 9.0).

Apparatus for Humidity Sensing

Schematic diagram of a relative humidity testing system is shown in Figure 1c. The sensor was installed in the testing chamber (9.42 L) which kept with the constant temperature at 26 °C. The controlled humidity environments were achieved using water injections into the chamber, which yielded the required humidity ranging from 8 to 95% (*Silica gel* desiccant was used to dry the air in the chamber for humidity testing down to about 8% RH after checking with a commercial *humidity* sensor. And silica gel was removed prior to the subsequent water injection). A fan was used to accelerate the evaporation of water droplet suspended at the tip of the needle. There was also a Thermo-Hygrometer (Testo 605-H1, Testo Ltd., Germany) in our system to monitor the RH in the chamber and the operating temperature of our sensors. The sensing properties of the sensors to humidity were examined by measuring the resonance frequency shifts of QCM which due to the additional mass loading.

Results and Discussion

Morphology of Sensing Membranes

FE-SEM images of CS fibrous membranes are shown in Figure 2a-b. It can be observed that the resultant fibrous membranes had a three-dimensional (3D) porous structure, in which tiny polymer nanowires fill the gaps between larger beads. These fibrous membranes with porous structure were reported to have much larger specific surface than that of flat films [18-19]. The nanowire diameter distribution of CS nanowebs is shown in Figure 2d. The major distribution region of nanowire diameters was 15-35 nm with an average diameter of 25 nm, which was one order of magnitude less than that of common electrospun fibers. The standard deviation of the wire diameters in nanowebs was 4 nm. Therefore, the specific surface of fibrous membranes was believed to increase with the formation of nanowebs. The formation of these webs was considered to be due to the fast phase separation of polymer and solvent during the charged droplet flight inside the high electric field [20] and the increased electrical conductivity of the polymer solution [21]. Additionally, even though the CS nanowebs were obtained at 2 wt% concentration, many beads as well as droplets were formed. This phenomenon was ascribed to the surface tension and viscoelastic properties of the solution [22]. Figure 2c shows the morphology of PEI-CS membranes. It can be seen that the PEI-CS membranes showed a relatively high surface roughness by forming some protuberances, which could be regarded as the irregular deposition of PEI membrane. One of the amazing advantages of this drop casting technique is the formation of a highly tailored polymer thin-film on the substrate, thus leaving the inherent high surface area of the electrospun membrane intact.

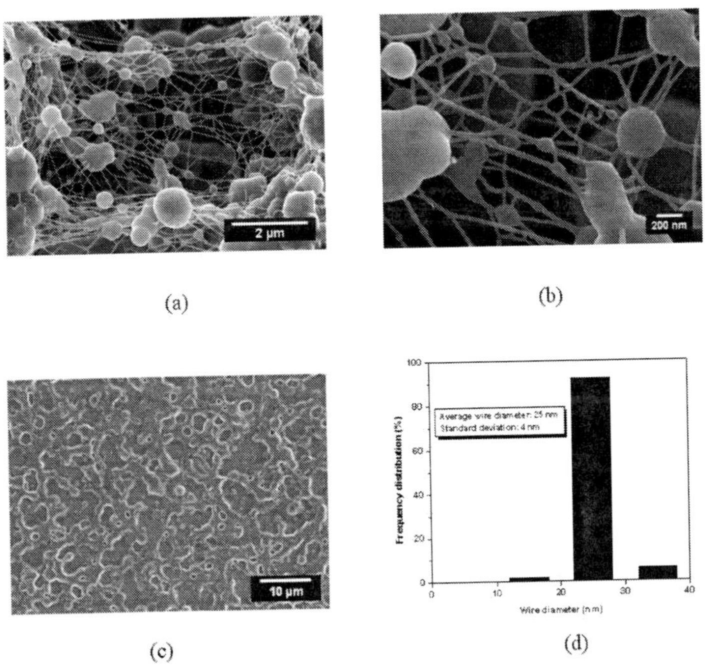

Figure 2. (a) Low and (b) high magnification FE-SEM images of CS fibrous membranes. (c) FE-SEM images of PEI-CS membranes. (d) Histogram showing the diameter distribution of nanowires shown in (a). (Reprinted with permission from [19]. © 2010 Freund Publishing House Ltd.).

Effect of the Type of Polyelectrolytes

Figure 3a shows the response curves of various polyelectrolytes coated CS membranes-based QCM sensors as a function of the relative humidity. The QCM sensor coated with fibrous CS membranes did not showed a visible change of frequency until humidity reached as high as 44% RH. The sensors showed a response with the maximum frequency shift of 278 Hz at 44% RH. The maximum frequency shifts of the fibrous CS membranes coated QCM sensors were 455, 810, 1174, and 1833 Hz at 61%, 80%, 90%, and 95% RH, respectively. It was observed that the response of sensors to RH was gradually increased with increasing the RH of the chamber.

The response of sensors to humidity was primarily ascribed to the hydrophilic hydroxyl groups ($-OH$) and amine groups ($-NH_2$) of CS, which can absorb moisture in the membranes easily. At 44% RH, the sensor coated with PAA-CS membranes began to present a relatively obvious decrease of frequency shifts. The sensors showed a response with the maximum frequency shift of 296 Hz and 2191 Hz at 44% and 95% RH, respectively. The hydrophilic carboxyl groups ($-COOH$) of PAA can absorb moisture in the membranes, and in turn lead to decrease of the frequency.

The PEI-CS membranes coated sensor presented the most obvious decrease of frequency shifts in the range of 8-95% RH. At the relative humidity of 95%, the maximum frequency shift of the sensor was 4022 Hz. The strong hydrophilic amine groups ($-NH_2$) of PEI can absorb moisture in the membranes easily. The frequency variations of the sensors showed nearly three orders of magnitude change during the RH variation of 8-95%, indicating the high humidity sensitivity. Compared with another two QCM sensors, the sensor coated with PEI-CS membranes showed the largest response. The fibrous membrane with higher specific surface area and larger porosity could absorb more water molecules to lead higher sensitivity (Figure 2c). From the sensing results, we can also conclude that PEI was more sensitive to humidity than PAA.

Besides, the three QCM sensors all showed a good linear relationship to the RH ranging from 8% to 70%. However, once the RH exceeds 70%, the frequency shifts increased in a steeper way. The reason for this is that the water molecules aggregated and formed clusters on the assembled fibrous membranes [23].

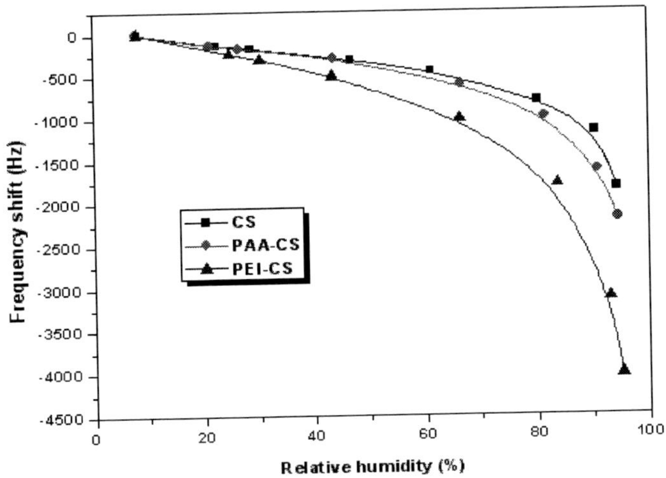

(a)

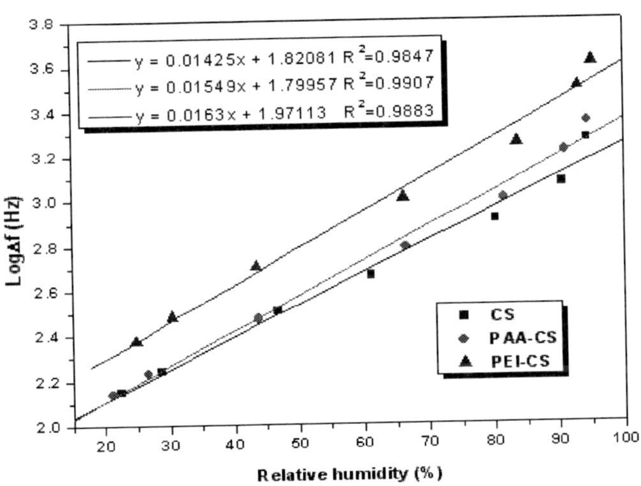

(b)

Figure 3. (a) Frequency shifts of various polyelectrolytes coated CS fibrous membranes-based QCM sensors as a function of the relative humidity.(b) Dependence of the frequency shifts of sensors on relative humidity (20-95%) and the calibration curves obtained from (a). (Coating load of CS fibrous membranes: 4000 Hz; Coating load of PAA or PEI deposited on CS fibrous membranes: 1200 Hz) (Reprinted with permission from [19]. © 2010 Freund Publishing House Ltd.).

Although the frequency shifts as a function of the relative humidity presented a nonlinear tendency with increasing RH, we can observe that the frequency shifts followed a logarithmic increase with increasing RH and all the QCM sensors exhibited good linear $Log(\Delta f)$ towards RH in the range of 20-95%. Linear plots of the $Log(\Delta f)$ as a function of relative humidity (20-95% RH) for three samples are shown in Figure 3b. The regression coefficients, R^2, were 0.9847, 0.9907, and 0.9883 for fibrous CS membranes, PAA-CS membranes, and PEI-CS membranes, respectively. The QCM sensors exhibited good linear $Log(\Delta f)$ to RH, indicating their potential application for highly sensitive humidity sensors.

Effect of the Coating Load of PEI

Figure 4 shows the response curves of PEI-CS coated QCM sensors as a function of the relative humidity with various coating loads of PEI. The sensor, with PEI coating load of 6000 Hz, presented the obvious decrease of frequency shifts in the range of 8-95% RH. At the relative humidity of 95%, the maximum frequency shift of the sensor was 13456 Hz, which was much higher than that (9185 Hz) of sensor with a PEI coating load of 3600 Hz. The frequency variations of the sensors showed nearly four orders of magnitude change during the RH variation of 8-95%, indicating the high humidity sensitivity. Additionally, both of the QCM sensors exhibited good linear $Log(\Delta f)$ towards RH in the range of 20-95%, as shown inset of Figure 4.

Effect of the Coating Load of CS

In order to know the effect of the coating load of CS fibrous membranes on the sensor sensitivity, various CS coating loads coated QCM sensors, on which the same content of PEI was deposited, were prepared. The sensing properties were also investigated, as shown in Figure 5. At 95% RH, the maximum frequency shift of PEI-CS membranes with a CS coating load of 5000 Hz was 10135 Hz, which was much higher than that (7271 Hz) of sensor with a CS coating load of 1200 Hz. The QCM sensors with a larger coating load exhibited a larger frequency response for the same RH since the absorption sites increased with increasing the coating load [17].

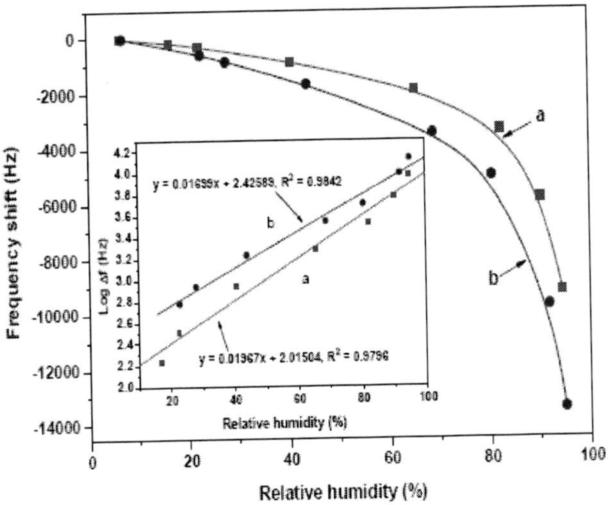

Figure 4. Frequency shifts of PEI-CS membranes coated QCM humidity sensors as a function of the relative humidity with various PEI coating loads (a) 3600 and (b) 6000 Hz. The inset is the linear logarithmic plot obtained from the same date (Coating load of CS fibrous membranes: 4000 Hz). (Reprinted with permission from [19]. © 2010 Freund Publishing House Ltd.).

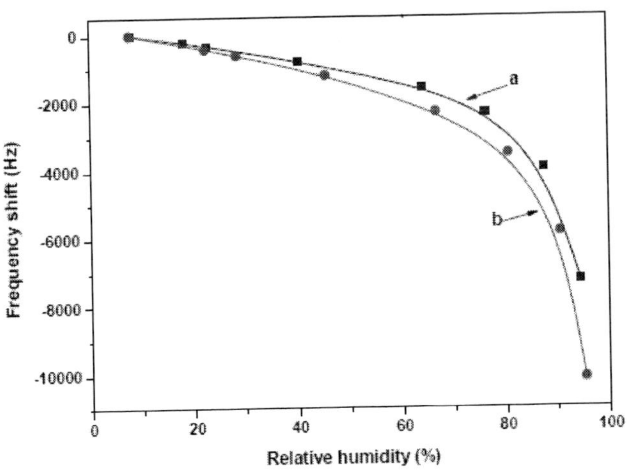

Figure 5. Frequency shifts of PEI-CS membranes coated QCM humidity sensors as a function of the relative humidity with various CS coating loads (a) 1200 and (b) 5000 Hz. (Coating load of PEI: 3600 Hz). (Reprinted with permission from [19]. © 2010 Freund Publishing House Ltd.).

Humidity Hysteresis and Response Time

The hysteresis and response time are crucial parameters for evaluating the performance of a humidity sensor. Figure 6 shows the humidity hysteresis characteristic of PEI-CS membranes coated QCM sensor. After reaching the equilibrium at 95% RH, the sensors were exposed to ambient air with RH of 70% until the full desorption was achieved.

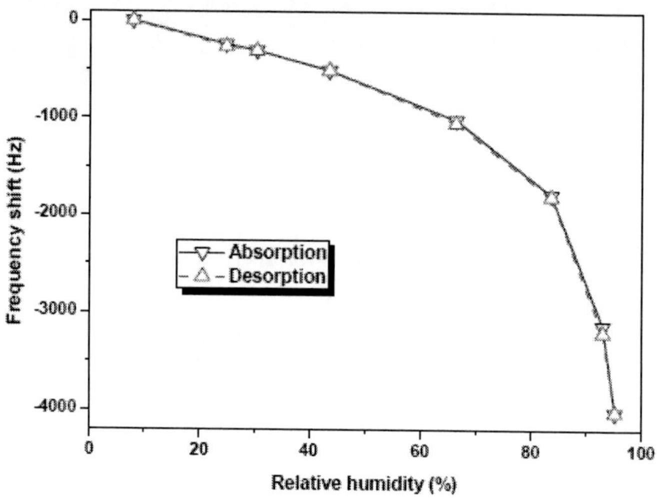

Figure 6. Humidity hysteresis characteristic of PEI-CS membranes coated QCM sensor. (Coating load of CS: 4000 Hz; Coating load of PEI: 1200 Hz). (Reprinted with permission from [19]. © 2010 Freund Publishing House Ltd.)

The solid line stands for the course from low to high RH, corresponding to the absorption process, while the dash line stands for the opposite direction, corresponding to the desorption process. It can be seen that the plots of frequency of the QCM sensor vs. humidity almost overlapped during humidification and desiccation processes, and the hysteresis was estimated to be as small as ~2%RH. Figure 7 shows the response curve of sensor coated with PEI-CS membranes with the RH changing from 8 to 95%. When the RH was increased from 8 to 95%, the response time (defined as the time required to reach 90% of the final equilibrium value) were 10, 20, 47, 64, 81, 120, and 300 s, respectively. Such a fast response behavior could be attributed to the 3D rough structure of PEI-CS membranes. Such structure can facilitate fast mass transfer of water molecules to and from the interaction region.

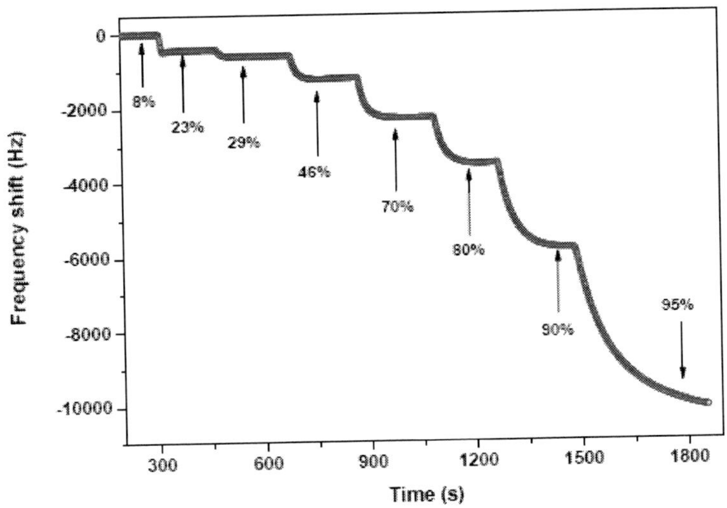

Figure 7. Response curves of PEI-CS membranes coated QCM humidity sensors with the relative humidity changing from 8% to 95%. (Coating load of CS: 5000 Hz; Coating load of PEI: 3600 Hz). (Reprinted with permission from [19]. © 2010 Freund Publishing House Ltd.)

Kinetic Analyses Using Modified Langmuir Model

The behaviors of absorption of water vapor molecules by PEI-CS membranes in various ranges of RH were compared. The reaction between the sensing membranes and the water vapor molecules is expressed simply as the follows:

$$\text{Sensing membranes} + \text{water molecules} \underset{k_{diss}}{\overset{k_{ass}}{\rightleftarrows}} \text{Sensing membranes—water molecules} \qquad (2)$$

where k_{ass} and k_{diss} are the rate constants for the adsorption and desorption processes, respectively. According to Langmuir adsorption isotherm model

[12, 24], the rate of surface reaction for forming a monolayer on the surface is described by Eq. (3)

$$\frac{d\theta}{dt} = k_{ass}(1-\theta)C - k_{diss}\theta \tag{3}$$

where θ represents the fraction of surface coverage, C is the water vapor concentration in the air, k_{ass} and k_{diss} are the rate constants for the adsorption and desorption processes, respectively.

Since QCM is a continuous, real time detector, it is possible to assess the kinetics of interaction. According to the Sauerbrey equation, that the decrease of frequency Δf is directly proportional to the attached mass of moisture molecules, and using $\Delta f_{\max}$ as the frequency change after a complete saturation of the surface of PEI-CS membranes with water vapor molecules. The relationship between the surface adsorption kinetics and Δf can be expressed as following:

$$\frac{d\Delta f}{dt} = -(k_{ass}C + k_{diss})\Delta f + k_{ass}C\Delta f_{\max} \tag{4}$$

During adsorption process, Δf is equal to $\Delta f_{\max}$ for very long time periods. Therefore, Eq. (4) is integrated to give:

$$\Delta f = \Delta f_{\max}\frac{k_{ass}C_{\max}}{k_{ass}C_{\max} + k_{diss}}(1 - e^{-(k_{ass}C_{\max}+k_{diss})t}) \tag{5}$$

where t is the time, $C_{\max}$ is the concentrations of water molecules in high RH conditions. The experimental data are compared with the data obtained from adsorption isotherm given in Eq. (5) with the k_{ass} and k_{diss} values obtained from the nonlinear least square fit for each curve versus time as shown in Figure 8.

The equilibrium constant (K) for PEI-CS membranes were calculated by the kinetic parameters data illustrated in Table 1 with the following equation:

$$K = \frac{k_{ass}}{k_{diss}} \tag{6}$$

when the RH was increased from 70 to 80%, from 80 to 90%, and from 90 to 95%, the values of equilibrium constant were calculated to be 3.533×10^5, 5.158×10^5 and 1.591×10^5 M^{-1}, respectively. Curve-fit equations of three absorption curves are given in Table 2.

The Gibbs free energy ΔG of adsorption process in terms of K at a constant temperature is expressed as [25]:

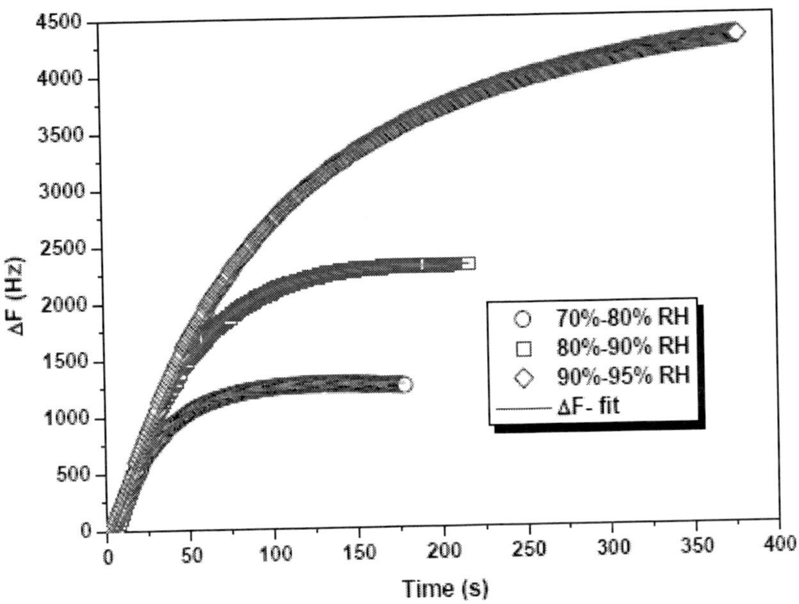

Figure 8. The responses curves intercepted from Figure 7 with the relative humidity changing from 70% to 95% (symbols) and the fit (red solid lines) to the Langmuir adsorption isotherm given in Eq. (5) with the k_{ass} and k_{diss} values obtained from the nonlinear least square fit for each curve versus time. (Reprinted with permission from [19]. © 2010 Freund Publishing House Ltd.).

Table 1. The adsorption and desorption kinetic parameters determined by using Langmuir model with least square fit of Eq. (5) to the experimental data given in Figure 8. (Reprinted with permission from [19]. © 2010 Freund Publishing House Ltd.)

RH (%)	Δf_{max} (Hz)	Concentration of vapor molar ($\times 10^{-4}$ M)	k_{ass} ($M^{-1}S^{-1}$)	k_{diss} ($\times 10^{-3} S^{-1}$)	K ($\times 10^{5}$ M^{-1})	Gibbs free energy, ΔG (kJ M^{-1})
70-80	1265	6.3	76.302	0.216	3.533	−31.704
80-90	2307	6.3	45.905	0.089	5.158	−32.643
90-95	4300	6.3	15.751	0.099	1.591	−29.724

Table 2. Curve-fit equation obtained from Figure 8. (Reprinted with permission from [19]. © 2010 Freund Publishing House Ltd.)

RH (%)	Curve-fit equations	R^2
70-80	y = 1259.36-2861.50/(1+exp((1.94+x)/20.71))	0.99737
80-90	y = 2314.13-6428.15/(1+exp((15.83+x)/34.47))	0.99847
90-95	y = 4443.06-85766.26/(1+exp((290.80+x)/99.77))	0.99897

$$\Delta G = -RT \ln K \qquad (7)$$

The Gibbs free energy for adsorption process is calculated using Eq. (7) for each curve and results are illustrated in Table 1. The Gibbs free energy for adsorption of three curves are calculated as −31.704, −32.643 and −29.724 kJ/mol, respectively. The negative value indicated that there is an energy loss for water molecules on PEI-CS membranes surface, since the water molecules are adsorbed into the PEI-CS membranes during adsorption process. This phenomenon consistent with the principle that the Gibbs free energy is negative ($\Delta G < 0$) for a spontaneous reaction. Therefore, the kinetic analysis for PEI-CS membranes shows a favored kinetic reaction with humidity molecules.

Conclusions

PAA-CS and PEI-CS membranes as sensitive coatings were successfully deposited on the electrode of QCM via electrospinning and subsequent drop casting to form a highly sensitive humidity sensor. The resultant sensors possessed good sensing characteristics such as high sensitivity, good $Log(\Delta f)$ linearity (20-95% RH), rapid response behavior, wide RH range of response, and small hysteresis. Increasing the coating load of the membranes introduced more absorption sites in the sensing membranes, and hence greatly enhanced the sensor sensitivity. When the RH was increased from 70 to 80%, from 80 to 90%, and from 90 to 95%, the values of equilibrium constant were calculated to be 3.533×105, 5.158×105 and 1.591×105 M^{-1}, respectively. The Gibbs free energy for adsorption of three curves are obtained as −31.704, −32.643 and −29.724 kJ/mol, respectively. The polyelectrolytes modified electrospun nanofibrous CS membranes have a great potential for humidity sensing applications at room temperature operations.

Acknowledgments

This work is supported by the National Natural Science Foundation of China (No. 50803009 and 10872048), the “111 Project” (No. 111-2-04 and B07024), the Shanghai Committee of Science and Technology (No. 10JC1400600), the National Basic Research Program of China (973 Program, 2011CB606100), the Innovation Program of Shanghai Municipal Education Commission (11ZZ59), and the Fundamental Research Funds for the Central Universities. This review collected some data from our previous publications of reference 19.

References

[1] A.E. Dessler and S.C. Sherwood, *Science* 323 (2009) 1020.
[2] Z.Y. Li, H.N. Zhang, W. Zheng and Y. Wei, *J. Am. Chem. Soc.* 130 (2008) 5036.
[3] X.F. Wang, B. Ding, J. Yu, M. Wang and F. Pan, *Nanotechnology* 21 (2010) 055502.
[4] B. Ding, J.H. Kim and S.M. Shiratori, *Sens. Actuator B* 101 (2004) 373.

[5] G. Sauerbrey, *Z. Phys* 155 (1959) 206.

[6] J.W. Grate, *Chem. Rev.* 100 (2000) 2627.

[7] F.W. Zeng, X.X. Liu, D. Diamond and K.T. Lau, *Sens. Actuator B* 143 (2010) 530.

[8] J.S. Bae and M.S. Gong, *Polymer-Korea* 20 (1996) 996.

[9] D. Verma and V. Dutta, *Sensor Lett.* 7 (2009) 143.

[10] P.G. Su and K.H. Cheng, *Sens. Actuator B* 142 (2009) 123.

[11] N.B. Cho, T.H. Lim and M.S. Gong, *Sens. Actuator B* 130 (2008) 594.

[12] S. Okur, M. Kus, F. Ozel and M. Yilmaz, *Sens. Actuator B* 145 (2010) 93.

[13] J.H. He, Y.Q. Wan and M.Y. Yu, *Int. J. Nonlinear Sci. Numer. Simul.* 5 (2004) 243.

[14] J.H. He, *Int. J. Mod. Phys. B* 22 (2008) 3487.

[15] J.H. He, Y.Q. Wan, L. Xu, *Chaos Solitons Fractals* 33 (2007) 26.

[16] X.F. Wang, B. Ding, M. Sun, J.Y. Yu and G. Sun, *Sens. Actuator B* 144 (2010) 11.

[17] B. Ding, M. Yamazaki and S. Shiratori, *Sens. Actuator B* 106 (2005) 477.

[18] B. Ding, C.R. Li, and S. Shiratori, *Nanotechnology* 17 (2006) 3685.

[19] X.F. Wang, B. Ding, J.Y. Yu, J.H. He, G. Sun, *Int. J. Nonlinear Sci. Numer. Simul.* 11 (2010) 509.

[20] M. Bognitzki, W. Czado, T. Frese, A. Schaper, M. Hellwig, M. Steinhart, A. Greiner and J.H. Wendorff, *Adv. Mater.* 13 (2001) 70.

[21] R.R. Yang, J.H. He, L. Xu and J.Y. Yu, *Polymer* 50 (2009) 5846.

[22] X.Y. Geng, O.H. Kwon and J.H. Jang, *Biomaterials* 26 (2005) 5427.

[23] E.R. Kleinfeld and G.S. Ferguson, *Chem. Mater.* 7 (1995) 2327.

[24] Y. Okahata, M. Kawase, K. Niikura, F. Ohtake, H. Furusawa and Y. Ebara, *Anal. Chem.* 70 (1998) 1288.

[25] S.J. Qiu, L.X. Sun, H.L. Chu, Y.J. Zou, F. Xu, N. Matsuda, *Thin Solid Films* 517 (2009) 2905.

In: Humidity Sensors
Editor: Christopher T. Okada

ISBN 978-1-61209-246-1

Chapter 5

MONITORING RELATIVE HUMIDITY USING ZNO BASED FILM BULK ACOUSTIC-WAVE RESONATOR

Xiaotun Qiu and Hongyu Yu

School of Electrical, Computer and Energy Engineering, Arizona State University, Tempe, AZ, 85287, USA

School of Earth and Space Exploration, Arizona State University, Tempe, AZ, 85287, USA

ABSTRACT

Humidity is a dynamic parameter that is essential for various fields of industry as well as human lives. There is a substantial interest in the development of relative humidity (RH) sensors for applications in monitoring moisture level at home, in clean rooms, cryogenic processes, medical and food science, and so on. Humidity sensors based on changes in the capacitance or resistance of the sensing element from absorption of water vapor have been investigated extensively. Alternatively, surface acoustic wave (SAW) resonant sensors have also been developed with polymer films deposited on top of a SAW resonator as the sensing layer. In this chapter, a novel RH sensing device using ZnO based film bulk acoustic-wave resonator (FBAR) was described. The resonant frequency of the FBAR decreased in a two-stage manner as RH increased in the environment. For low RH (RH < 50%), a frequency downshift of 2.2 kHz per 1% RH change was observed. This effect was attributed to water molecules replacing the adsorbed oxygen on the ZnO surface, thus increasing the density of the film, resulting in a frequency drop. For high RH (RH > 50%), a frequency downshift of 8.5 kHz per 1% RH change was obtained, which was due to the mass loading effect of the water layers formed on the ZnO surface. It was demonstrated that the two-stage response of the FBAR can be interpreted using the power law theory for semiconductor gas sensors and the mass loading effect of the resonator. UV light was applied to monitor its effects on the humidity sensing performance of the FBAR. UV can enhance the sensitivity at low RH (frequency downshift increased to 3.4 kHz per 1% RH change), while degrading the sensitivity at high RH (frequency downshift decreased to 5.7 kHz per 1% RH change). The mechanism of the influence of the UV illumination was interpreted.

INTRODUCTION

Relative humidity (RH) is a very common parameter in our environment. It is essential not only for human comfort, but also for a broad spectrum of industries and technologies. There is a substantial interest in the development of RH sensors for applications in monitoring moisture level at home, in clean rooms, cryogenic processes, medical and food science, and so on. Table 1 summarizes the widespread applications of RH sensors.

Table 1. Applications of relative humidity sensors

Applications	Relative Humidity Range (%)
Air conditioner	40~70
Microwave oven	2~100
Incubator	50~80
Dehydrated food processing	0~50
Electric device manufacturing	0~50
Hygrometer	0~100

The requirements that RH sensors need to meet for various applications are as follows: (1) a good sensitivity over a wide range of relative humidity and temperature, (2) a short response time, (3) good reproducibility and small hysteresis, (4) good durability and long life time, (5) resistance against contaminants, (6) negligible temperature dependence and (7) low cost [1]. Today, roughly 75% of the RH sensors in the market are based on the capacitive principle [2]. Meanwhile, there are plenty of other RH sensors, such as resistive sensors [3] and surface acoustic sensors [4]. Capacitive sensors are based on dielectric parameter change of thin films upon moisture uptaking. The properties of these sensors depend on the film material and the electrode geometry. On the other hand, resistive sensors can transfer the RH variation into an impedance change, which can be measure by a current, a voltage or a resistance. Another alternative is monitoring RH with gravimetric effect using surface acoustic wave resonators (SAW). Thin plates of piezoelectric material were coated with a hygroscopic layer (usually a polymer layer). The resonant frequency of the SAW shifted as water was absorbed onto the surface. Thus, the RH can be measured via the frequency change.

In this chapter, a novel RH sensing device using ZnO based film bulk acoustic-wave resonator (FBAR) was introduced [5, 6]. The chapter is arranged in the following manner. The second part is the introduction of film bulk acoustic-wave resonator. The third part focuses on the design and the testing setup of the FBAR RH sensor. The fourth part consists of the experiment results and the explaination of the sensing mechanism. In the fifth part, the influence of the ultraviolet (UV) light on the humidity sensing performance of the FBAR is investigated. The last part is a shotr summary of the chapter.

FILM BULK ACOUSTIC-WAVE RESONATOR

Film bulk acoustic-wave resonator usually consists of a sputtered piezoelectric thin film (ZnO/AlN) sandwiched by two metal layers (top and bottom electrodes). A resonance condition occurs if the thickness of piezoelectric thin film (d) is equal to an integer multiple of a half of the acoustic wavelength (λ). The fundamental resonant frequency ($f=v / \lambda$) is then inversely proportional to the thickness of the piezoelectric material used, and is equal to $v/2d$, where v is the acoustic velocity at the resonant frequency in the piezoelectric thin film. Figure 1 shows the schematic structure of a FBAR.

Lakin *et al.* reported the first FBAR in the early 80's [7]. It was composed of a thin film of ZnO deposited on top of a thin silicon membrane supporting structure. The fundamental resonant frequency was near 500 MHz. Since then, FBAR has been developed both as filters [8] and as high sensitivity sensors [9, 10]. For sensing applications, Benetti *et al.* used FBAR deposited with Pd and Co-tetra-phenyl-porphyrin films to detect hydrogen, CO and ethanol [11]. Zhang *et al.* investigated Au coated FBAR as an Hg^{+} ion sensor in water [12].

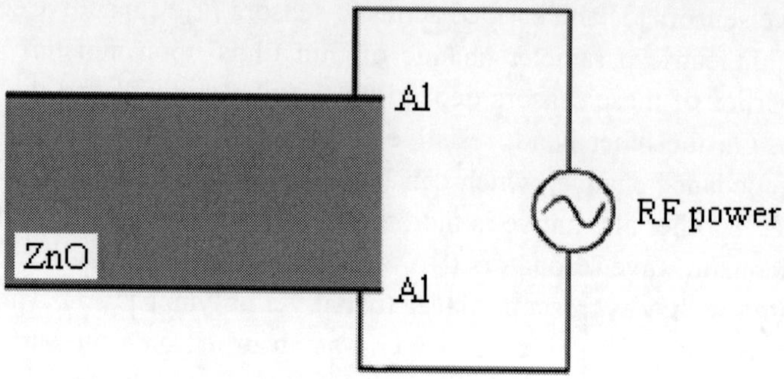

Figure 1. Schematic structure of a FBAR.

Lin *et al.* explored TNT (Trinitrotoluene) detector by introducing antibodies to the FBAR [13]. Yan *et al.* employed FBAR as a biosensor through biospecies coating [14]. All the sensors mentioned above adopted a similar sensing mechanism: mass loading effect.The resonant frequency of the FBAR decreased linearly with the mass accumulated on top of the resonator [9]. Different coatings were applied for the absorption of the sensing target. In this way, the mass accumulated on the resonator increased, resulting in a decrease of the resonant frequency, which was used as a detecting signal for the FBAR sensor. Another promising application is using FBAR as an ultraviolet sensor [10]. The resonant frequency varied due to the density change of the ZnO film induced by UV illumination.

FBAR RH SENSOR

The schematic structure of the FBAR RH sensor is shown in Figure 2. The FBAR was fabricated on top of a SiN diaphragm (0.6 μm thick). A sputtered ZnO film (1.2 μm) acted both as the RH sensitive layer and the piezoelectric actuation layer for the FBAR sensor. The ZnO film was characterized by X-ray diffraction (XRD) using Cu K_α radiation (Figure 2). Only the Bragg reflection corresponding to (002) planes was observed, indicating that the film had preferred orientation along the wurtzite C axis, normal to the silicon substrate. The top and bottom electrodes were made of Au (0.2 μm) and Al (0.2 μm), respectively. The fabrication process of the FBAR RH sensor was as follows. In the first step, a SiN layer was deposited on a Si wafer (100) with low-pressure chemical vapor deposition (LPCVD). Then the Si wafer was etched from the backside anisotropically in potassium hydroxide (KOH) to form the cavity. Next, the bottom Al electrode was deposited by electron-beam (e-beam) evaporation and patterned on top of the SiN film using wet chemical etching. ZnO was radio-frequency (RF) sputtered and etched to form the desired pattern. The last step was the e-beam deposition and patterning of top Au electrode by lift-off.

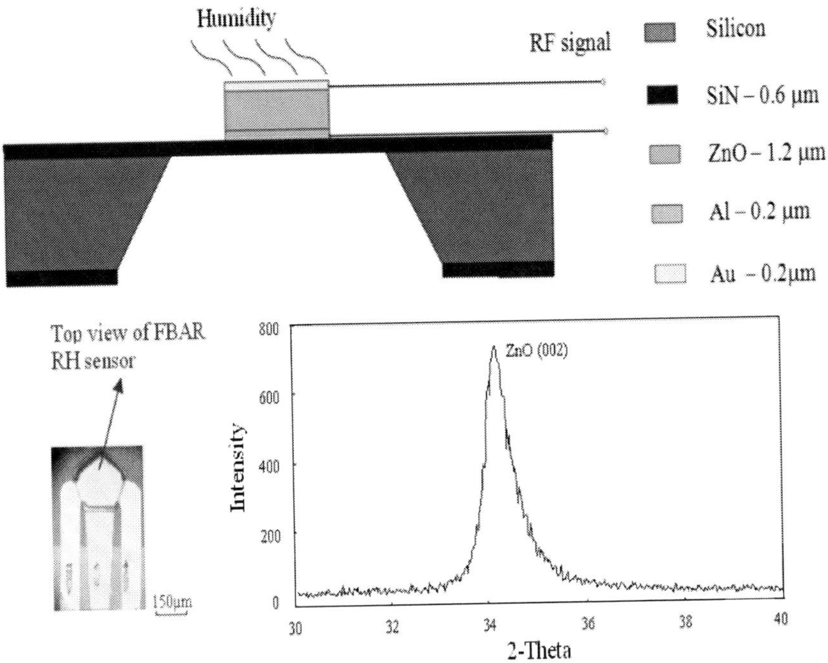

Figure 2 Schematic cross-sectional structure (not drawn to scale) of the FBAR RH sensor with a photograph of the top view of a fabricated device on the left and an XRD trace of the ZnO film illustrating that it had <002>crystal orientation on the right.

The sensor was encapsulated in a chamber, in which RH was controlled by an ultrasonic humidifier and measured by a HH314A humidity temperature meter. The resonant frequency of the FBAR was monitored with an Agilent E5071C network analyzer and recorded by a LabVIEW program. It had a noise floor of 0.7 ppm at 1.4 GHz, which was suitable for integration with a wireless sensor network. The quality factor (Q) of the FBAR was between 530 and 550. A versatile hand-held ultraviolet lamp (365 nm) was used as the UV source. A UVX digital ultraviolet intensity meter was applied to calibrate the UV power received by the FBAR.

EXPERIMENT RESULTS

The RH response of the FBAR sensor at room temperature is shown in Figure 3. A two-stage process can be identified based on the different slope values. At low RH (RH<50%), the resonant frequency decreased linearly with the RH and a frequency downshift of 2.2 kHz per 1% RH change was observed. With the current noise floor, the detection limit was around 0.45% RH. The RH response in this range was due to the replacement of adsorbed oxygen with water molecules on the ZnO surface [15]. Therefore, the density of the ZnO film increased. The resonant frequency of the FBAR can be determined from the following two equations: $v=(E/\rho)^{1/2}$ and $f=v/2d$, where E, ρ and d are the elastic constant, density and thickness of the ZnO film, respectively. v is the acoustic velocity within the ZnO film and f is the resonant frequency of the FBAR [10]. As the density of the ZnO film increased with RH, the acoustic velocity decreased, resulting in the decrease of the resonant frequency.

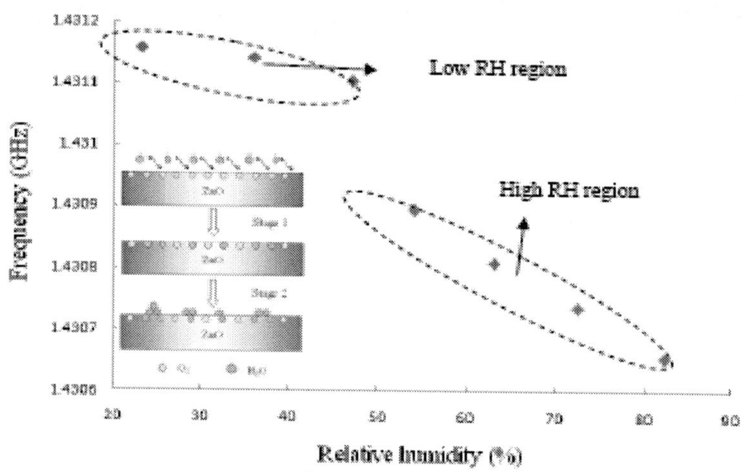

Figure 3 The RH response of the FBAR at room temperature. Two stages can be identified. At low RH (RH<50%), a frequency downshift of 2.2 kHz per 1% RH change was observed. At high RH (RH>50%), a frequency downshift of 8.5 kHz per 1% RH change was obtained. The inset illustrated the mechanism of the two-stage response of the FBAR. At low RH (stage 1), the response was due to the replacement of adsorbed oxygen with water molecules on the ZnO surface. At high RH (stage 2), a discrete water layer began to form on the ZnO surface, which acted as a mass loading on the FBAR.

It has been known that the electric resistance of a semiconductor gas sensor exposed to a target gas (partial pressure *P*) is proportional to P^n where *n* is a constant fairly specific to the kind of target gas (power law). Yamazoe *et al.* [16] established a theoretical basis for the power law combining a depletion theory of the semiconductor, which deals with the distribution of electrons between surface state and bulk, with the dynamics of adsorption and/or reactions of gases on the surface, which is responsible for accumulation or reduction of surface charges. By extending their work to address the density change of the semiconductor gas sensor, a theoretical model for the RH response of the ZnO film based FBAR was developed for the low RH region (RH<50%).

According to Yamazoe *et al.* [16], the density of conduction electrons at the surface [*e*] can be expressed as:

$$[e] = N_d \exp(-\frac{m^2}{2}) \tag{1}$$

where N_d is the density of donors and *m* is the reduced depletion depth. Using the double Schottky barrier model, the amount of adsorption/desorption of oxygen at each grain-boundary is proportional to [*e*]. Thus the relative density change, $\Delta\rho$, induced by adsorption/desorption of oxygen is proportional to [*e*].

Meanwhile, the frequency shift Δf can be formulated as:

$$\Delta f = \frac{1}{2d}\left(\sqrt{\frac{E}{\rho+\Delta\rho}} - \sqrt{\frac{E}{\rho}}\right) = \frac{\sqrt{E}}{2d}\frac{\left(1-\sqrt{1+\frac{\Delta\rho}{\rho}}\right)}{\sqrt{\rho+\Delta\rho}} \tag{2}$$

In the equation above, $\rho \gg \Delta\rho$. In this way,

$$\Delta f \approx \frac{\sqrt{E}}{2d}\cdot\frac{\left(1-1-\frac{1}{2}\frac{\Delta\rho}{\rho}\right)}{\sqrt{\rho}} = -\frac{\sqrt{E}}{4d\rho\sqrt{\rho}}\Delta\rho \tag{3}$$

Therefore the frequency shift Δf is proportional to $\Delta\rho$ and thus, it is also proportional to [*e*].

At room temperature, oxygen is adsorbed as O^-_2 on the ZnO surface [17].

$$O_2 + e \underset{k_{-1}}{\overset{k_1}{\rightleftarrows}} O_2^- \tag{4}$$

Here e stands for a conduction electron at the surface, and k_1 and k_{-1} are the rate constants of forward and reverse reactions, respectively. The rate of accumulation of O^-_2 is given by

$$\frac{d[O_2^-]}{dt} = k_1 PO_2[e] - k_{-1}[O_2^-] \tag{5}$$

Here the brackets mean the density of O^-_2 per unit area or that of electrons per unit volume, t is time, and PO_2 is the partial pressure of oxygen. At equilibrium, the rate is zero, thus,

$$k_{o1} PO_2[e] = [O_2^-] \tag{6}$$

$$k_{o1} = \frac{k_1}{k_{-1}} \tag{7}$$

where k_{o1} is the equilibrium constant of oxygen adsorption.

Assuming that there are no electron-trapping sites other than O^-_2 on the surface, then O^-_2 ions would be solely responsible for the surface charge density, that is,

$$[O_2^-] = N_d \omega \tag{8}$$

Here ω is the depletion depth and $\omega = mL_D$, where L_D is the Debye length [16].

Insert (1) and (8) into (6) yields,

$$k_{o1} PO_2 \exp(-\frac{m^2}{2}) = \omega \tag{9}$$

As mentioned above, water molecules will take the place of the adsorbed oxygen on the ZnO surface [15]. Two water molecules can dissociate and become chemisorbed in the ZnO film [15, 18]. At equilibrium,

$$k_{o1}PO_2[e] = [O_2^-] + k_{o2}PH_2O[O_2^-] \tag{10}$$

$$k_{o2} = \frac{k_2}{k_{-1}} \tag{11}$$

where k_2 is the rate constant of water absorption and PH_2O is the partial pressure of water vapor, which is proportional to RH at a certain temperature.

From the discussion above, the power law exponent n, defined as $d(\log\Delta f)/d(\log PH_2O)$, is obtained,

$$\frac{d(\log \Delta f)}{d(\log PH_2O)} = 1 - \frac{1}{1+m^2} \tag{12}$$

where m is sufficiently large under usual conditions [16]. Thus n is around 1, which can be used to explain the linear relationship between the resonant frequency and the RH in Figure 3 in the low RH region.

In Figure 3, at high RH (RH>50%), the resonant frequency decreased linearly with the RH and a frequency downshift of 8.5 kHz per 1% RH change was obtained, corresponding to a detection limit of around 0.12% RH. With increasing RH, a discrete water layer began to form on the ZnO surface, which acted as a mass loading on the FBAR. Thus the resonant frequency of the FBAR decreased linearly with the mass of the water accumulated on top of the resonator [9]. This can also be observed from the Q response of the FBAR with increasing RH (Figure 4(a)). At RH higher than 50%, Q decreased quickly with humidity, while at RH lower than 50%, Q changed less than 10%. These results indicated that water layer formed on the ZnO surface when RH was higher than 50%, thus attenuating acoustic wave, resulting in a lower Q. Figure 4(b) shows the photographs of the FBAR under different RH. It was clear that at RH higher than 50% (P3) there were small water drops formed on the device.

The Butterworth-Van Dyke (BVD) equivalent circuit of FBAR with and without liquid loading is shown in Figure 5, where C_o, L_m, C_m, and R_m are the clamped capacitance between the top and bottom electrodes, motional inductance, motional capacitance, and motional resistance of the resonator, respectively. R_2 and L_2 are associated with acoustic energy loss and mass loading by the liquid, respectively [9].

(a)

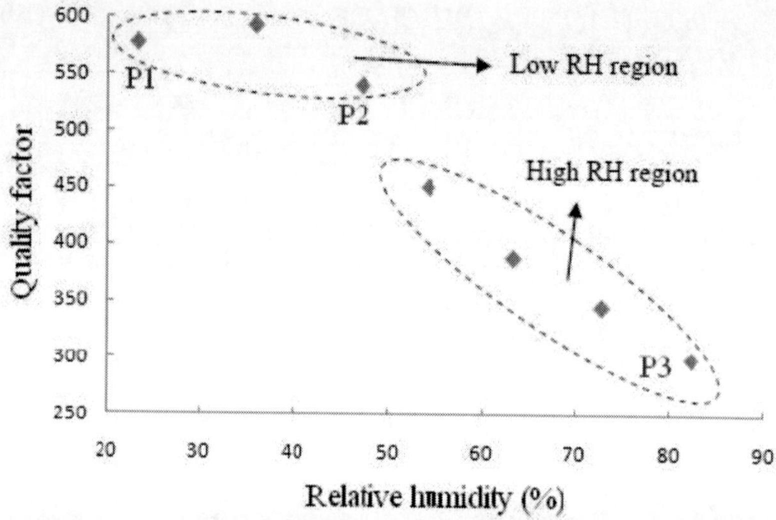

(b)

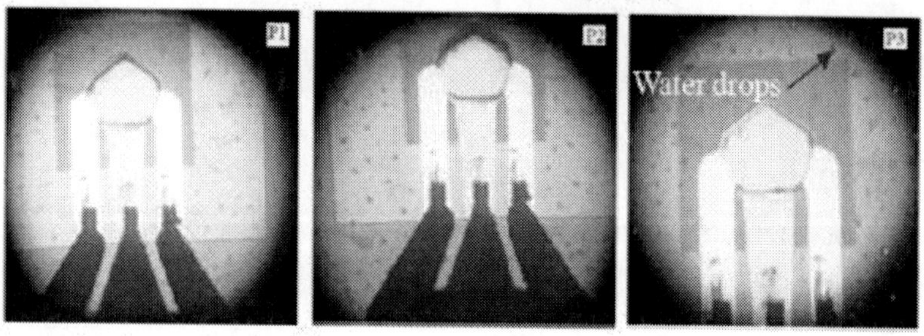

Figure 4 (a) The response of FBAR's Q vs. RH. Two different stages can be identified. At RH lower than 50%, Q changed less than 10%. However, at RH higher than 50%, Q decreased quickly with humidity. (b) The photographs of the FBAR under different RH. P1: 23.2%; P2: 47.2%; P3: 82.3%. The corresponding positions of these three humidity values are indicated on the figure of the Q response of the FBAR sensor in (a). It was clear that at RH around 82.3% there were small water drops condensed on the device.

In the equivalent circuit, the circuit elements are related to the physical parameters of the resonator and the added liquid by

$$f_s = \frac{1}{2\pi\sqrt{L_m C_m}} \tag{13}$$

$$f_p = \frac{1}{2\pi\sqrt{L_m \frac{C_m C_o}{C_m + C_o}}} \tag{14}$$

$$L_m = \frac{1}{4\pi^2 f_p{}^2 C'} \tag{15}$$

$$C' = \frac{C_m C_o}{C_m + C_o} \; (\tag{16}$$

$$f'_p = \frac{1}{2\pi\sqrt{(L_m + \Delta L)\frac{C_m C_o}{C_m + C_o}}} \tag{17}$$

$$\Delta L = L_2 \tag{18}$$

$$L_2 = \frac{2L_m \rho_1 d_1}{\rho_0 d_0} \tag{19}$$

where f_s and f_p are the series and parallel resonant frequency of the FBAR. f'_p is the parallel resonant frequency with the liquid loading. ρ_0, d_0, ρ_1, and d_1 are the density of the piezoelectric layer, the thickness of the piezoelectric layer, the density of the added liquid and the thickness of the added liquid, respectively [9].

From the equations above (similar results can be obtained for f_s),

$$\Delta L = \frac{1}{4\pi^2 C'} \frac{f_p^2 - f'^2_p}{f_p^2 f'^2_p} \approx \frac{1}{4\pi^2 C'} \frac{2\Delta f_p}{f_p f'^2_p} \tag{20}$$

$$\Delta f_p = f_p - f'_p \tag{21}$$

$$\frac{\Delta f_p}{f_p} = \frac{\Delta L}{2} \frac{1}{L_m + \Delta L} \approx \frac{\Delta L}{2L_m} = \frac{\rho_1 d_1}{\rho_0 d_0} \tag{22}$$

Equation (22) indicates that the resonant frequency shift of an acoustic resonator is linearly related to the mass of a material added on the top of the resonator. Assuming the mass of the water accumulated on the FBAR is proportional to the RH, a linear response of the FBAR can be expected in the high RH region, which is in agreement with the experiment results.

Influence of Ultraviolet on the Performance of FBAR RH Sensor

Previous study showed that the adsorption of water on the ZnO surface can be enhanced with ultraviolet illumination [19]. Thus UV may improve the RH response of the ZnO based FBAR. The RH response of the FBAR under UV (600 $\mu W/cm^2$) is shown in Figure 6. It was also a two-stage response. At low RH (RH<50%), a frequency downshift of 3.4 kHz per 1% RH change was observed with a detection limit of 0.3% RH. It was demonstrated that in this region UV can enhance the adsorption of water on the ZnO surface, resulting in a higher sensitivity. While at high RH (RH>50%), a frequency downshift of 5.7 kHz per 1% RH change was obtained, corresponding to a detection limit of around 0.18% RH. This value was degraded compared with the detection limit in the UV absent case (around 0.12% RH). It can be explained as follows: when there is UV illumination on the ZnO film, the wettability of the surface will be enhanced, resulting in a more hydrophilic surface [19-23]. Therefore under UV illumination, due to the improved hydrophilicity, the water layer on the FBAR distributed more uniformly (Figure 6) with smoother surface and less mass loading effect compared with the UV absent case, where the ZnO surface was less hydrophilic. The frequency shift of the FBAR due to additional mass loading can be estimated from equation (22). With a smoother surface, the effective thickness of the water layer on the ZnO film under UV illumination was smaller compared with the UV absent case, resulting in a reduced frequency shift. Thus, the RH response under UV was smaller in this region. This was in agreement with the measured Q response of RH under UV illumination (Figure 7). Q decreased more slowly at high RH in this case, due to the reduced attenuation of acoustic wave with the smoother water layer.

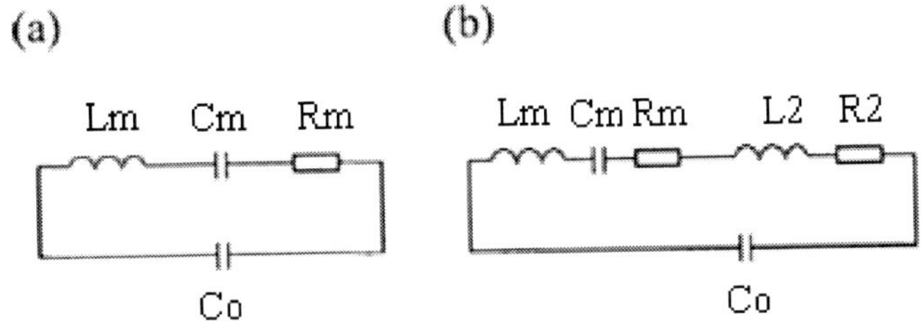

Figure 5. Butterworth-Van Dyke (BVD) equivalent circuits: (a) a FBAR without liquid loading; (b) a FBAR with liquid loading.

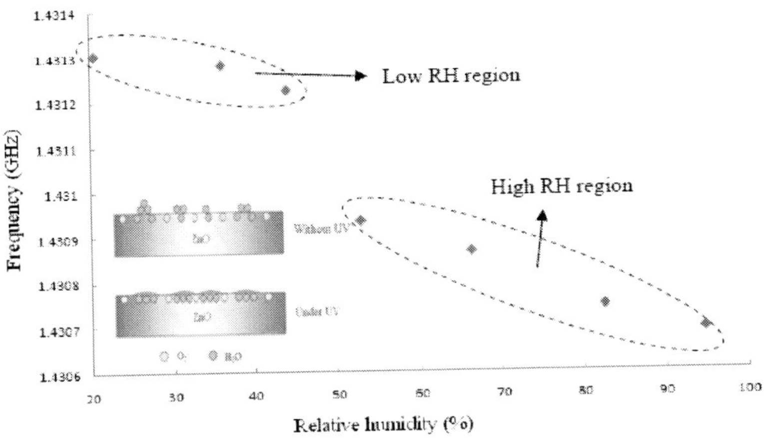

Figure 6. The RH response of the FBAR under UV (600 μW/cm^2) illumination. It was a two-stage response. At low RH (RH<50%), a frequency downshift of 3.4 kHz per 1% RH change was observed. It was demonstrated that in this region UV can enhance the adsorption of water on the ZnO surface, resulting in a higher sensitivity. While at high RH (RH>50%), a frequency downshift of 5.7 kHz per 1% RH change was obtained. This value was smaller compared with the UV absent case. The inset showed the response of the FBAR sensor at high RH region in the presence and absence of UV illumination. When there is UV illumination on the ZnO film, the wettability of the surface will be enhanced, resulting in a more hydrophilic surface. Therefore under UV illumination, due to the improved hydrophilicity, the water layer on the FBAR distributed more uniformly with smoother surface and less mass loading effect compared with the UV absent case, where the ZnO surface was less hydrophilic.

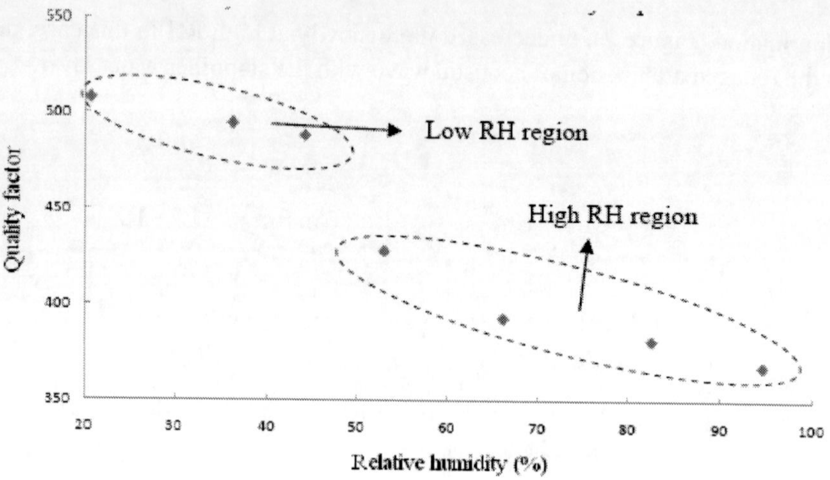

Figure 7 The response of FBAR's Q vs. relative humidity under UV (600 μW/cm^2) illumination. At RH higher than 50%, Q decreased more slowly compared with the UV absent case.

SUMMARY

In summary, a RH sensor was developed using ZnO based FBAR. The resonant frequency of the FBAR decreased linearly in a two-stage manner as the RH increased in the environment. For low RH (RH<50%), a frequency downshift of 2.2 kHz per 1% RH change was observed. This effect was attributed to water molecules replacing the adsorbed oxygen on the ZnO surface, thus increasing the density of the film. While for high RH (RH>50%), a frequency downshift of 8.5 kHz per 1% RH change was obtained, which was due to the mass loading effect of the water layers formed on the ZnO surface. UV light was applied to monitor its effects on the humidity sensing performance of the FBAR. UV can enhance the sensitivity at low RH (frequency downshift increased to 3.4 kHz per 1% RH change), while degrading the sensitivity at high RH (frequency downshift decreased to 5.7 kHz per 1% RH change). The mechanisms of both the RH response and the influence of UV illumination were investigated. This study has proven the feasibility of monitoring RH with ZnO film based FBAR.

REFERENCES

[1] Z. M. Rittersma. Sens. *Actuators*. B. 2002, vol 96, 196-210.
[2] G. Scholz. *Technisches. Messen*. 1992, vol 59, 88-109.
[3] P. M. Faia; C. S. Furtado; A. *J. Ferreira. Sens. Actuators*. B. 2004, vol 101, 183-190.
[4] M. Penza; G. Cassano. *Sens. Actuators*. B. 2000, vol 68, 300-306.
[5] X. Qiu; J. Oiler; J. Zhu; Z. Wang; R. Tang; C. Yu; H. Yu. *Electrochemical and Solid-State Letters*. 2010, vol 13, J65-J67.
[6] X. Qiu; R. Tang; J Zhu; J. Oiler; C. Yu; Z. Wang; H. Yu. *Sens. Actuators*. B. 2010, vol 147, 381-384.
[7] K. M. Lakin; J. S. Wang. *Appl. Phys. Lett.* 1981, vol 38, 125-127.
[8] M. Ueda; M. Hara; S. Taniguchi; T. Yokoyama; T. Nishihara; K. Hashimoto; Y. Satoh. *Jpn. J. Appl. Phys.* 2008, vol 47, 4007-4010.
[9] H. Zhang; E. S. Kim. J. Microelectromech. Syst. 2005, vol 14, 699-706.
[10] X. Qiu; J. Zhu; J. Oiler; C. Yu; Z. Wang; H. Yu. *Appl. Phys. Lett.* 2009, vol 94, 151917.
[11] M. Benetti; D. Cannatà; F. Di Pietrantonio; V. Foglietti; E. Verona. *Appl. Phys. Lett.* 2005, vol 87, 173504.
[12] H. Zhang; M. S. Marma; E. S. Kim; C. E. McKenna; M. E. Thompson. IEEE Trans. Ultrason. Ferroelect. Freq. Contr. 2007, vol 54, 1723-1725.
[13] A. Lin; H. Yu; M. S. Waters; E. S. Kim; S. D. Goodman. *Proceedings of IEEE MEMS* 2008. 2008, 208-211.
[14] Z. Yan; X. Y. Zhou; G. K. H. Pang; T. Zhang; W. L. Liu; J. G. Cheng; Z. T. Song; S. L. Feng; L. H. Lai; J. Z. Chen; Y. Wang. *Appl. Phys. Lett.* 2007, vol 90, 143503.
[15] Y. Li; F. Valle; M. Simonnet; I. Yamada; J. Delaunay. *Appl. Phys. Lett.* 2009, vol 94, 023110.
[16] N. Yamazoe; K. Shimanoe. Sens. Actuators. B. 2008, vol 128, 566-573.
[17] M. Takata; D. Tsubone; H. Yanagida. *J. Am. Ceram. Soc.* 1976, vol **59**, 4–8.
[18] G. Korotchenkov; V. Brynzari; S. Dmitriev. Sens. Actuators. B. 1999, vol 54, 197-201.
[19] R. Sun; A. Nakajima; A. Fujishima; T. Watanabe; K. Hashimoto. *J. Phys. Chem.* B. 2001, vol 105, 1984-1990.
[20] X. J. Feng; L. Feng; M. H. Jin; J. Zhai; L. Jiang; D. B. Zhu. J. *Am. Chem. Soc.* 2004, vol 126, 62-63.
[21] H. Liu; L. Feng; J. Zhai; L. Jiang; D. B. Zhu. *Langmuir*. 2004, vol 20, 5659-5661.

[22] J. Han; W. Gao. *J. Electron. Mater.* 2009, vol 38, 601-608.
[23] M. Miyauchi; A. Shimai; Y. Tsuru. *J. Phys. Chem.* B. 2005, vol 109, 13307-13311.

INDEX

A

B

C

D

E

F

G

H

N

O

P

Q

R

S

T

U

V

W

X

Y

Z